AF556883

Rosaluna

The Sustainable World

Aims and Objectives

Sustainability is a key concept of 21st century planning in that it broadly determines the ability of the current generation to use resources and live a lifestyle without compromising the ability of future generations to do the same. Sustainability affects our environment, economics, security, resources, health, economics, transport and information decisions strategy. It also encompasses decision making, from the highest administrative office, to the basic community level. It is planned that this Book Series will cover many of these aspects across a range of topical fields for the greater appreciation and understanding of all those involved in researching or implementing sustainability projects in their field of work.

Topics

Data Analysis
Data Mining Methodologies
Risk Management
Brownfield Development
Landscaping and Visual Impact Studies
Public Health Issues
Environmental and Urban Monitoring
Waste Management
Energy Use and Conservation
Institutional, Legal and Economic Issues
Education
Visual Impact
Simulation Systems
Forecasting
Infrastructure and Maintenance
Mobility and Accessibility
Strategy and Development Studies
Environment Pollution and Control
Land Use
Transport, Traffic and Integration
City, Urban and Industrial Planning
The Community and Urban Living
Public Safety and Security
Global Trends

Main Editor

Rosaluna

Enzo Tiezzi
University of Siena, Italy

Enzo Tiezzi
University of Siena, Italy

Cover: Illustration adapted from an old ceramic of Vietri (Naples).

Published by

WIT Press
Ashurst Lodge, Ashurst, Southampton, SO40 7AA, UK
Tel: 44 (0) 238 029 3223; Fax: 44 (0) 238 029 2853
E-Mail: witpress@witpress.com
http://www.witpress.com

For USA, Canada and Mexico

WIT Press
25 Bridge Street, Billerica, MA 01821, USA
Tel: 978 667 5841; Fax: 978 667 7582
E-Mail: infousa@witpress.com
http://www.witpress.com

British Library Cataloguing-in-Publication Data

A Catalogue record for this book is available
from the British Library

ISBN: 978-1-84564-399-7
ISSN: 1476-9581

Library of Congress Catalog Card Number: 2009930793

Printed in Great Britain by Athenaeum Press Ltd.

Contents

Yucatán
canal de Yucatán
Cuba
Mérida
Tulum
Bacalar
Palenque
Tikal
Turneffe il.
Blue Hole
Mexico
Livingstone
Quiriguá
Guatemala
C. Guatemala
Tegucigalpa
S. Salvador
Nicaragua

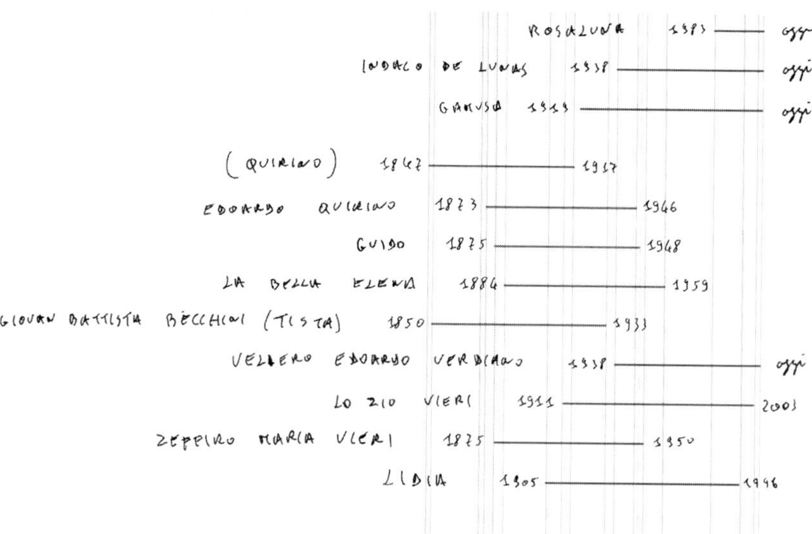

Chronology of real and imaginary characters

1
Hot "manatee" springs

I Bagni Marìi

Rosaluna was born in the hot springs. She became acquainted with the reassuring warmth of the spring waters when still in her mother's womb. She waited until the moon changed, and on mid-winter's day, under the sign of Aquarius, she began to kick to announce her arrival. Her mother was in the only natural pond of hot water near Rapolano, in the woods of the Bagni Marìi. The pond was circular and about two and a half metres in diameter. On the cold winter's day, the hot vapours formed a dense white cloud over the pond. The air had a strong but basically reassuring smell of boiled eggs. Rosaluna calculated that coming out in the warm water would be less traumatic for both of them and began to kick insistently. Marianna felt the contractions begin and relaxed into the sensation.

Rosaluna was born underwater in the manner of dugongs and manatees, our distant mammal cousins. She saw an opaque liquid and smelt the rotten-egg gas, emerging to draw in her first breath and emit her first cry. The vapours wafted the perfume of pine resin and the amber moon peeped down through the pine fronds.

Many years later Rosaluna remembered these sensations when young Jemal grasped her around the waist and lifted her out of the hot waters. They made love in a pond in a Hungarian wood in the fading spring light, imagining the paintings of Csontváry, the cabalist artist from Pécs.

Perhaps opening her eyes in the sulphur water, perhaps her slightly cross-eyed gaze when she grew up had something to do with the fact that Rosaluna saw two moons, or two half moons, or two sickles, or two black pearls at new moon. She saw them with a pale ray that reflected on the sea surface. You can do it too, if you stare intensely at the moon with slightly wet eyes, eyes burning with salt or thermal water, like those of Rosaluna. Stare at the moon or the sun for a long time with your eyes closed and you will see Goethe's colours: vermilion, carmine, ochre. To see two moons, more concentration is needed. It is a scientific phenomenon, the opposite of Gaumont's and the Lumière brothers' stereoscope. In the stereoscope there is a photographic plate exposed from two points seven centimetres apart, the distance between your eyes, and the double plate gives you the sense of depth.

Rosaluna's grandfather, Veliero Edoardo Verdiano, had a 1899 Gaumont stereoscope in mahogany and brass. He bought it from an old antique dealer in Budapest and the apparatus came with a desk that closed like a harmonium and contained 100 little drawers in mahogany and brass, each containing 40 plates of views of Europe from that period: lifesavers wearing striped costumes, the first scandalous undines in bathing costume, and velocipedes.

When she was 11, Rosaluna saw two moons for the first time and exclaimed, "Oh!" Then she remembered a scene in the haystack of the farm in the Chianti region of Siena. She realised for the first time that she could go back in time, to 50 years earlier. At first she was afraid: she knew that it was impossible to go back in time. Then she felt bewildered that she could let herself wander in time. Her third sensation was an attitude of listening and great humility.

Anna Borgianni cut herself a shady cranny in the haystack under the August sun. She had the healthy smell of an adolescent with budding breasts and desires. She was the daughter of a God-fearing communist family. A moment of fragrance and sensations.

Rosaluna observed the scene with the same moist eyes with which she looked at the moon, the two moons. The stories wafted confusedly into and out of her imagination. At 16, she read *The Thousand and*

One Nights and things became less confused, more transparent. All she had to do was open a door and walk into other stories, love stories, adventure stories, family stories, travel stories, stories of your contrada. One story contains another, like the stories of Scheherazade, and all of them come from the "treasures of Nature that go back to the first light," according to the Peruvian poet, Carlos Germán Belli.

Rosaluna had a grandfather called Veliero Edoardo Verdiano, an uncle-in-law, Uncle Vieri, a great-grandfather engineer, Edoardo Quirino, and a great-great-grandfather pharmacist, Giovan Battista, *Tista* for short. The women in the family were all in her. Rosaluna conversed with these ancestors every day. She listened to them, or not; it depended. They had lived longer than she and the exchange of information was continuous and fertile.

Thus, Rosaluna found herself at the crossroads of unimaginable and imaginable stories, which create new stories by treasuring old ones and making room for imagination, with a love for traditions and things to be conserved, with a dreamy creative gaze towards the future. The three main stories of that crossroad were a bourgeois story that began in the Chianti hills and set out for oases, deserts and Islam; the second was an alchemical story beginning between Mt. Amiata and the Rapolano springs and proceeding on the Guatemalan plateau; the third was revolutionary, beginning in St. Louis, Missouri, developing between Siena and the sustainable world.

The Mediterranean tornado

The sea birds had only awoken about half an hour ago but they decided to return to roost. They hid in the most inaccessible crevices, sheltered from both the maestrale and scirocco. Michele closed the stable door with the big bar and lay down among his donkeys, partly to comfort them and partly to comfort himself. At least it was warm there. The clouds had the wildest forms. They seemed riderless horses funnelling through the San Martino curve in the Palio of Siena.

Nobel laureate Prigogine had predicted it. The greenhouse effect increased evaporation of the Mediterranean and the increase in temperature and masses of water vapour created turbulence. The resulting dissipative structures no longer obeyed linear laws. Chaos and cyclones like those of the Caribbean were possible. In the Mediterranean Sea, however, birds, plants, people and animals have no experience of these events. They are defenceless and do not know what to do. The only rational reaction is fear.

As she accelerated the sage green Audi, Rosaluna sensed that this was no normal storm. The old tufo and stone house had been there for more than a thousand years, the great ground-floor rooms could shelter people and animals as well as the car. It might even be advisable to shelter in the cellars carved in the tufo. She called Giovanna on her mobile phone and agreed on a plan.

That evening horses, ducks, doves, hens, peacocks and dogs sheltered in the house. The great ex-convent had become a Noah's ark. Only ten kilometres from Pacina, Rosaluna saw the black centre of the vortex approaching from behind. The Audi responded well on the curves and up the Pietraia rise, and finally entered the Massa road that led to the farm.

Giovanna was waiting with the garage open. All were safe in the belly of the large house. The double grey doors of the garage closed behind the two women. Rosaluna jumped out of the car, hugged Giovanna, and the two of them went through the narrow passage to join the others in the cellar.

In the days that followed, the men and women of the Chianti set to work to saw up the uprooted olives, to set tractors back on their wheels, to clean roofs and shelters. A plough had ended up in one of the cypresses at l'Apparita, the same tree that presided over Giulia's wild lovemaking. This was the very story that Teresina, the old maid of the great-grandmother, told during the night spent in the cellars of Pacina.

Giulia from l'Apparita

At the time of our story, Teresina was 110 years old. Round and wide like a Russian matrioska, she didn't show her years. She seemed perennially on the verge of tipping between vertical and horizontal position.

Have you ever tried to spin an egg in transverse position? After a while, the couples of friction predominate and the egg spins on its tip. Who knows if this was Christopher Columbus's trick? Today the experiment is published in the *Physical Review*.

According to Teresina, Giulia was young, beautiful and hot blooded, and her bust was never far from view. This may have been due to low necklines or to the fact that her dresses of black percale used to wear thin in this region. The story was also about Gilleno who claimed to work in the "building art." For years he had driven many types of cars (that did not belong to him) without a licence. He had been an active black market trader during the war and had done the odd year of prison for sexual molestation, though the charge was never proved. He got out by an amnesty and became expert in the same field, his consultancy earning him enough to buy a German sidecar. His expertise in amnesties of various kinds brought many thefts to fruition in pre-amnesty periods. In later years, still a good-looking man, his fame had spread from the lower Valdichiana as far as the Chianti.

Giulia was fascinated by his bad reputation, but most of all she was pleasantly overwhelmed by his manner of making love. Unbeknown to his wife, Gilleno took Giulia in his sidecar to dance at Montecatini Terme. With her high heels, Giulia became the immediate postwar boogie-woogie queen of Montecatini, and also danced the tango and milonga. She invented all possible excuses to go to the village to shop: "I forgot the bread at the bakery;" "They said to come back for fresh eggs;" "I left the salame on the butcher's counter;" . . . . Rosaluna's great-grandmother closed an eye (or both eyes): in those days bread was baked in the farm oven, eggs came from the hen house and the salame and buristo of black Siena hogs made at Pacina at Epiphany was the best in the area.

The village was a kilometre and a half away, but Giulia actually went "shopping" at the cypresses of l'Apparita, a kilometre from home, and she disappeared into the bushes with Gilleno on the slopes of the olive grove. She was lucky that she never conceived. It is said that she eventually set up house with a Welsh officer, becoming his faithful wife and a model mother. On a windy day at l'Apparita, the cypresses pretend never to have seen that splendid youthful body, nipples to the wind, rollicking among the bushes and flocks of starlings.

Indaco de Lunas

Rosaluna's first love was the young Turk, Jemal. Their story is too beautiful to tell. Rosaluna's platonic lover, on the other hand, was Indaco de Lunas. He was a cartoon character, but for her he was one of the family and followed her everywhere, even when she went back in time.

After two bankruptcies, when he lost all his savings and the money of some friends whom he enticed into his business ventures, Indaco decided to invest in serious things. This decision was the fruit of experience. The first business failure taught him not to bet on stupid animals. "Pigeons," he said, "see everything and understand nothing." Thus, the prize-winning carrier-pigeon farm at Berardenga was a complete failure. Instead of carrying messages where they were meant to go, the birds flew to the feed trough. Indaco was vague about the second business venture. "That Bali white was a great fighter. He left the ring with the enemy beaten," Indaco would say, failing to mention the $10,000 lost on bets and the more than $20,000 that went up in smoke for the prize-winning Berardenga fighting-cock farm.

The truth was that that evening, in a Bali smoking house, Indaco and his dear friend Simone Cannavacciuolo, a young merchant of great lineage and rare wisdom, embraced and made an appointment in Piazza del Campo in Siena. The idea was to produce a tufo wine from Sangiovese and Syrah within the walls of Siena. Syrah stirs the thoughts, like young Cuban prostitutes. Pilar, Indaco said, is a young Cuban whore. She is clean and ingenuous like a natural pearl. Therefore, Pilar is not a whore, but a pearl of a girl. The problem, according to Indaco, was that no-one knows where Syrah comes from. Enological scholars can tell you its genealogy and the original structure can be discovered by magnetic resonance, but before that? What is its true story?

It is a story of random encounters and of choices, from the times of the Crusades down to today. It also has to do with Cardinal Branda Castiglioni, between Lake Balaton and Vezprem, and his master winemakers and the Benedictine abbey with one of the most beautiful libraries in the world. Syrah is like amber that despite itself contains a fossilised orchid. Whores are like amber; they contain the orchid of sex, which is different from the flower of love. Orchids, wine vapours, women's hair and amber all have fractal structure: beautiful configurations, never the same, rich in biodiversity, and the smoke rings and puffs of Cuban cigars are the same.

Indaco de Lunas and his friend Simone Cannavacciuolo subsequently made a lot of money with Syrah, Guatemalan jade, Burmese lacquerware and Kashmiri moonstones, or at least this was the good epilogue of Indaco's paper story. However, in this case, the paper story and the true story had the same ending. Many years later, when Indaco had grown wise with the help of Rosaluna and Giovanna, he invested much of his money to improve the production of Syrah and Sangiovese at Pacina. He restored an old farmhouse on the banks of the Malena, inspired by drawings of Gaudi, and so as not to belie himself completely, had a house built in a large oak tree. Here he would occasionally retire to read poems of Giovanni Pascoli in summer and Kahlil Gibran in spring, when "the soul unfolds in a thousand petals like the lotus." Gabriele D'Annunzio kept him company on autumn afternoons.

When he needed money, Indaco sent a cable to faithful Abdul in Srinagar, who brought the semiprecious stones he traded in Kerala, to Siena. The stones were inlaid with gold in an original and very tasteful manner by Bettina Martini di Cigala, a dear family friend, and sold in the Tuscan markets.

True and paper stories criss-crossed and Rosaluna tripped from story to story, as happens to anyone who dreams or lets the imagination wander.

The two-tailed mermaid

Galenas and Old Laces

Here time has become entropy
among galenas and old laces,
the desert its triumph celebrates
with dunes gilded by ancient south-west winds.

Here the mines temples have become
new Petra in the land of Sardinia
but art got lost with the craft
and time no more the due information brings.

No more the railway carries blendes
and precious galenas to the beach of Piscinas,
sailing-wheel-boats by steam drawn
would wait for an everlasting time at the old mole.

If the mermaid had not had two tails, we could not have made love and the dugong would have remained in my dreams, the dream of Indaco de Lunas, a paper character. The woman in the red dress insisted with the same old story, that I was a character created by the pencil of a cartoonist and that I existed only in the imagination of my creator. But the woman who was telling me this, to what world did she belong? She was so rational in trying to convince me that I wasn't real! But if she, too, was speaking to me, she must have been an imaginary character as well and her rationality was merely mist or dream.

This was no easy problem and my scientist friend and I had discussed it for nights on end: is it dreams or is it the illusion of reason that generates monsters? We were inclined to favour the second proposition, despite our fondness of dreams.

That September morning, the sea at Piscinas di Ingurtosu in Sardinian island was a calm indigo. Indaco de Lunas had woken unusually early and remembered the details of a long dream: the cartoonist, the pastel shades of his watercolours and the story of William Saroyan recounted by his grandmother. He pinched his thigh to be sure he was awake, opened the window over the sea, taking a deep breath of iodine and salty air, donned his swimming costume and set out across the wide beach towards the shore. It is true that the scene was set for magic meetings, that the Sibyl had indicated his animal totem as the dugong, and that after the Great Greenhouse Cycle, marine mammals from the warm oceans were even entering the Mediterranean, but the encounter that morning had something incredible about it.

But let us return to reality. The scene was the ex-mine of Ingurtosu and Naracauli, with its Stratford-on-Avon-style mansion of Lord Brassey and the blende and galena hopper, down to the dunes by the sea. The prize-winning railway Naracauli–Piscinas, built in 1871 to take the minerals down to the wharf where they were loaded on sailing ships and steamers, is now only a simulacrum with rails reaching into voids, as in the film by Fitzcarraldo, without even a uniformed guard to defend its memory. The great golden beach has mixed the stories.

The dugong met Indaco 11 metres from the shore. It brushed him, dancing elegantly despite its awkward size, and with two pirouettes, invited him to follow. Indaco pinched his left calf, washed his eyes better with seawater and followed the dugong. When the colour of the water faded from indigo to turquoise and a ray of September sun lit the depths of the sea, she appeared in all her beauty. Her nipples were slightly dark, like inverted convolvuli, her long hair the colour of Guatemala amber, her buttocks muscular. She could have been 20 or even 40. She was beautiful and it was the first time that Indaco had seen a mermaid with two tails. They entwined for long hours among the waves. It was she who moved with the agility of a sea creature.

In the weary afterglow, the mermaid said, "You see, Indaco, I cannot follow you to shore because the sea is my world and you are a cartoon character."

"Not so," replied Indaco, "I pinched myself twice. You, the two-tailed mermaid, are the imaginary character!"

"Did you enjoy making love with me?"

"Enormously!"

"Well, that means you are dreaming, because you think I am an imaginary mermaid."

They conversed at length of poetry and dreams, art and cartoons, they kissed and made love again, discovering a sense of wonder. Now they knew that *this* was real and that they dreamed in other dimensions.

The two-tailed mermaid left Indaco, saying, "I'll be yours on land too. My name will be Egle and I will be wearing turquoise percale."

On reaching the shore, Indaco found his young scientist friend sitting between the tidelines and exclaimed, "Rediscover wonder!" Finally even scientific questions were clear. Their hearts rose to the high plain. In nature everything is true, they agreed, and the real delirium is to dream of dominating nature.

Many years later, Rosaluna had Indaco tell her the story of the two-tailed mermaid. Despite an age difference of almost half a century, he remains her dearest friend.

2

Treasures and horses from the Camargue to the Rif Mountains

The Legend of the Marsh Gypsy

When his eyes followed
the dark cloud
and left behind
the immense plain,
his gypsy glance
embraced the marsh
and he started sloping down
with a rude and fast step.

He went on walking
following his own instinct,
he caught a glimpse
of a landscape painted
by indigo and turquoise:
he feared no more
and abandoned to a dream
of love on the fourth sura.

In seventeenfifty
anno domini
it was easy to believe

that love could win:
he flew on the hippogriff
and in padule appeared
the circular pieve
with its ivy walls.

Came towards him a friar
sharp and black eyes;
stopped the gypsy
absorbed in his thoughts.
Fra Benedetto – he said –
more than friar, brother,
from the east I was awaiting for you
here I have a ring for you.

Two daughters were born
both of a moony beauty
the former studied logics
dreamt of the sea the latter.
Skilled in cards
was the first girl
in the cradle danced the other
an African dance.
This story goes on
surely it's not finished
its weft is subtle
and by nature warped.

At the caravanserai of Fez

For at least a fortnight, there had been no room for men, horses or dromedaries at the caravanserai of Fez, though the date of the horse auction was not for another week. In those days of 2001, the market of Fez, the wool dying quarters and the stalls with fossils from the Atlas mountains were mobbed by the auction participants, curious tourists often willing to spend some money to take something back to Europe, the Americas or Asia, while awaiting the real event, the horse to be won in the hotly contested auction.

Exactly a year earlier, at the start of the new millennium, Rosaluna's grandfather had received a strange visit in the old stone house in the Siena Chianti region. An old Citroen, the glorious admiral of the French automobile producer known as the "shark," pulled up, disgorging a beautiful young gypsy girl accompanied by three very old gentlemen, also dressed in precious silks and brocades in the manner of gypsies. The three gypsies introduced the girls as the queen elected at the start of the new millennium by their community in the Camargue, during a great meeting of Romany people. The three were the Council of Elders. They claimed to be more than a hundred years old and introduced themselves as Gaspar, Melchior and Baldassar, because they were bearing gifts. The girl had no name: she was just the queen.

Rosaluna's grandfather made the guests comfortable in the garden among flowers in various tones of violet and antique rose, offering them his best tea.

"It isn't from your area," he said, "it's golden broken orange pekoe from Assam, the best I found on my travels in Ceylon, India and Burma." After tea, the old gypsies took three chests of fragrant camphor wood with elegant studs and strange locks out of their *barrakans*.

"These," said Gaspar, "open only to the vibrations of your voice, only yours. They were designed according to figures that scientists call *strange attractors* that can be simulated with Prigogine's non-linear equations. They are vocalprint locks that respond to your vocalprint which is unique, like your fingerprint. Start talking and they will open."

The Tuscan gentleman drew deeply on his Havana cigar. When he was young, he looked like Che Guevara. His admiration for Che and his love for the aroma of Cuban cigars had never waned. He sipped a tiny glass of ratafia, which out of consideration he did not offer his guests, and began to recite *Alexandros* by Giovanni Pascoli. At the second last verse, where Alexander weeps, the three locks opened in unison.

If it hadn't been for that encounter the previous year, thought Veliero Edoardo Verdiano, today he would not have been with his granddaughter Rosaluna at the caravanserai of Fez. The great walled area, built outside the city, still had the features of a primitive inn: four main parts connected two by two at right angles and the internal courtyard packed with animals, usually dromedaries and mules, divided by various palisades. The stables for Arabian horses were in the buildings. "Horses are treated like gentlemen," thought Veliero, remembering the enthusiastic exclamations of the three old gypsies when he had shown them the stable built for the carriage horses in his Chianti house. It had been designed by Fantastici in the 19th century, with 11 monolithic travertine columns, 3 orders of symmetry in the curvature of the ceiling and walls, and 4 troughs, one for each horse, in yellow *broccatello* marble of the Siena Montagnola.

The caravanserai was shaded by great cedars. There was water in a semicircular trough with coloured *azulejos* on the bottom. The guest rooms had verandahs, indicating that the caravanserai, there in the extreme west of Morocco, had been built by a Persian master who had organised water storage as only Persians knew how.

The horses to be auctioned came from the Rif and Atlas mountains, as well as from many Arab countries. Unaccountably, the Saudi princes had brought many fine animals that year. Rich Moroccans from Tangiers, Meknes and Casablanca passed the time at *tavli*. Mauritanians, descendants of the main warrior tribes, the Beni-Hassan, Reguibat, Uled-bu-sba and Uled-Delim, were dressed in wide *draa* or *darraa* of sky-blue calico. Unarmed marabouts stood to one side, smoking *nargileh* or eating pistachios. Veliero noticed that an elderly Beni-Hassan, with a thick grey beard, was wearing a splendid aquamarine of about 30 carats. The sight of it reminded him of the opening of the chests brought by Gaspar, Melchior and Baldassar.

The first contained gold and silver: heavy bracelets in silver from Nubia, golden Shiite beads, some small gold idols with ruby eyes, others in gold inlaid with ivory or ebony, thousands of golden coins – marengos, old sterling, ancient Spanish doubloons. The second contained ambers and jades. Rosaluna's grandfather was a connoisseur and immediately realised the rarity and quality of the larger pieces of amber: they contained Devonian fossils, a small orchid, insects that became extinct millions of years ago, certainly from San Domingo. There were particularly fine ambers from the Baltic: a collection of the Lega Anseatica, precious gold and amber jewellery from Tallinn and Bremen, necklaces that must have belonged to a tzarina. The greatest surprise, however, was the jades. There were idols in dark green, almost black, jade and pale opalescent jade from Guatemala; there were transparent Chinese jades of inestimable value and violet jades from Burma.

The contents of the first two chests had dazzled Veliero Edoardo Verdiano and he could not understand the reason for the visit and his relationship with the opening of the locks.

"The vocalprint has genetic characteristics. Only the great-grandson of our queen of the year 1000 could open the locks," said Melchior, prostrating himself, and adding, "You know that your family descends from ancient generations of gypsies who settled the marshes of Valdichiana in the year 1000. They carried on our trades: tinkering, basket weaving, raising of horses." Rosaluna's grandfather nodded. "It is written," interrupted Baldassar, "that the treasure of the queen of the year 1000 goes to the first great-grandson at the start of the second millennium, one reason being that it was collected, together with other wealth, thanks to the alchemical inventions of the father of our queen of the year 1000. You are the one!"

Although the Tuscan gentleman was well off, he had never seen riches of these proportions. The third chest was the richest of all. It contained semiprecious stones of great quality and size (turquoises, moonstones, topazes, amethysts, lapis lazuli) and two Kashmiri papier mache' boxes, one full of rubies and the other of emeralds. A third, larger box, finely painted with miniatures, contained a thousand splendidly bright star sapphires. With a handful of these, thought Veliero, I can make my granddaughter's dream come true, by buying her the Arabian of her dreams at the horse auction in Fez.

The phone rang and Veliero went to answer it: it was his friend Beverlyfrom London, with the information on Andalusia and eating places in Cadiz that he needed. While he was speaking, he heard the motor of the Citroen and on returning to the garden, found only the open chests and the perfume of incense and myrrh. The four gypsies had disappeared and all that the old Tuscan gentleman could see was the cloud of dust rising from the cypress lined road to l'Apparita. Only then did he notice the aquamarines in the Burmese lacquer box, pendants with great droplets and other cut in rectangular form. One was an intense azure, like the one worn by the Beni-Hassan.

Altair

The sun reflecting from the aquamarine of the Mauritian warrior awoke Veliero from his memories. The Beni-Hassan was moving fast, followed by four sky-blue warriors. Rosaluna's grandfather caught the Arab words pronounced as they passed: someone is going to try to capture Altair!

"What did they say?" asked Rosaluna, seeing his worried glance.

"After them, Rosaluna! Someone wants to steal Altair."

The girl jumped to her feet, her eyes fiery, her hand on her great-great-grandfather's Winchester carbine. They followed the five Mauritanians to the stables on the opposite site of the courtyard. Rosaluna immediately noticed a cloud of dust rising outside the walls. As Rosaluna and Veliero left the enclosure to investigate the cause of the dust, a bulldozer that had demolished the wall with the clear intent of stealing the horses, the Beni-Hassan unsheathed two pistols and went into Altair's stable. Five or six people with cords of various lengths were trying to capture the splendid animal, with little success. Outside the hole, an agitated American was expostulating in English that they were good-for-nothings, to get a move on and that he had paid them too much. Veliero immediately recognised Mr. Lewis, a Miami millionaire, and couldn't help thinking that a theft by poor devils, like Arsenio Lupin or Robin Hood, was excusable, but that a rich man should try such a tactic was inconceivable. Mr. Lewis evidently intended to have the horse without taking part in the auction.

Rosaluna was pointing her loaded Winchester. Veliero had loaded two cartridges in his finely engraved Liege dog-lock flintlock gun. Mr. Lewis's men found themselves between two lines of fire and fled. The American magnate was apprehended by the local gendarmes, still expostulating about friends in high places, even in Morocco, and about lack of respect towards his person, etc. etc.

From that moment, Veliero Edoardo Verdiano and the Beni-Hassan became good friends. A few days later the Tuscan gentleman doubled the figure offered by a sheikh from Meknes for Altair, and obtained the horse. Rosaluna was wild with joy.

Altair was a large slender white horse, pure Arabian. He would have suited Amedeo Guillet, the cavalry officer of the Royal Italian Army, Commander of the Amhara Band on horseback in East Africa. The horse could be recognised instantly by the golden star-like mark on its forehead. It was no normal mark for a horse, especially a white horse. During the visit of the three old gypsies, Melchior had mentioned it. "In Arabia," he said, "a horse descended from the unicorns has been born. It does not have a golden horn, but like the panda's thumb, it has a sign recalling the biological evolution of the species: a golden star on its forehead."

The desert libraries

The grandfather, the Beni-Hassan and Rosaluna were sipping hibiscus at a playing table finely inlaid with ebony, ivory and mother-of-pearl. The Beni-Hassan was telling them about the oldest university in the world, the University of Cairo, founded in the year 1000, more than 100 years before the University of Bologna and "Two hundred and forty years before that of Siena," added Rosaluna.

"Our oldest and most precious libraries with illuminated manuscripts," said the Beni-Hassan, who had willingly accepted a cigar from Rosaluna's grandfather, "are the nomad libraries of the desert. They are passed from family to family to be studied and are kept in clay niches."

A young man came up to the table and handed a letter to Veliero Edoardo Verdiano, who glanced at the name of the sender. "Speak of the devil!" he exclaimed. He read out the letter-reportage received from their common friend, Rosario Simone:

> *Ten Labbe is a tiny oasis in the Adrar region of northern Mauritania. A sign welcomes the visitor; "We are almost there . . . in a few minutes we'll be in Ouadane!" After hours of driving in the desert, the taxi driver leaves the wheel to the passenger next to him to light a cigarette. When we reach Ouadane it is pitch dark, because the oasis, like most of the country, is without electricity. Ouadane means* two wadi; wadi *are the beds of ephemeral streams. "And where are the two wadi?" we ask one of the two boys who shows us to the village next morning. "One is the Wadi en-Nakhil which means stream of palms and the other is the Wadi el-Eilm which means the stream of knowledge," replies the boy. To reach the rock spur on which the ancient caravan city is built, we go through a flourishing palm grove which is the first wadi. In order to get an idea of the stream of knowledge, it is necessary to go to the house of Sidi Ould Abidine. The owner welcomes us with composed courtesy, swatting on a mat of palm leaves. Books belonging to his extended family are spread around him on the mat. Abidine proudly shows them to us: they are ancient texts that reached Ouadane centuries ago on camel back. "There are manuscripts of the Koran with side notes, editions of the Hadith which are traditions of the prophet Mohammed, astronomy, geography, Muslim law books . . . ," he explains. He is wearing an ample* darraa, *the typical, ironed and starched, sky-blue tunic. His old house is built of clay with acacia and palm wood. On the walls he has attached cuttings of articles and interviews that concern him, published in the specialist press of half of Europe. "Unfortunately," he adds, "these books are at the mercy of three sworn enemies: termites that become aggressive during the rare rains, exposure to light, and the dry climate." As he speaks, he shows us an ancient text that has been devoured by termites, offering us a magnifying glass at the same time.*
>
> *The old town of Ouadane is a small maze of alleys and dry stone walls. In one of these alleys is the tiny library, directed by Abidine, the jealous custodians of which are two stout black women. Most of the older houses are unlived in, but old people sitting outside tiny doors often invite strangers to drink tea. The houses are of the Arab type, with a large inner courtyard in which the family keeps a Bedouin tent. In every room there is a niche for books. It is not easy to find a means of leaving Ouadane. The only way*

of getting to Chinguetti is a rattly pick-up that takes the splendid carrots of the oasis to the market in Atar. The oasis of Chinguetti is situated at the end of the steppe, where high dunes of finest sand begin. Today, most of the alleys in the old part have been invaded by sand and dunes up to 20 metres high loom menacingly over the last houses. It is difficult to imagine that this oasis, a crossroads on the salt route, was a major religious and cultural centre from the 13th to the 16th century. All that remains is about fifteen private libraries and only one of the thirteen original mosques. The dry stone minaret towers above the other buildings. The most important book collections are those of the Habbut family, that of Ahmad Mahmoud, and those of Ahmad Sharif, the Audaa and the Abdel Hamid families. Our tour begins with one of the smallest, belonging to Sheikh Ould Bouhia. One of his sons ushers us in. The collection consists largely of 16th century texts of the Koran with verse by verse explanations. We then visit the collection closest to the old mosque, which belongs to Ahmad Mahmoud.

"The UNESCO project is ambitious on paper," Seif Eddine, the library director begins, as he binds his head in a black gauze houli. *Then he dips his calligraphy pen in one of his inkwells, ex-machine-gun shells, and continues, "Unfortunately, this and other projects never get off the ground and we cannot wait. The libraries of the oasis are all private and we, like Ahmad Mahmoud, have put together collections of five families to pool our efforts. Here there are about 400 texts of inestimable value, including medicine and astronomy books from Egypt, Syria, Iraq and other distant countries. The most valuable volume is a Koran about a thousand years old." As Seif Eddine speaks, some girls and a boy of the oasis are silently consulting some of the texts, each in a corner of the inner court of the old house. In order to gain access to the Habbut library, we go to the house of the librarian, Muhammed Ould Koulame. A boy shows us in and a woman of the family, swathed in a colourful* melhafa, *invites us to drink tea, offering us dates to dip in a bowl of butter. With its 1500-odd manuscripts, the Habbut library is the largest in Chinguetti.*

"The author," added the grandfather, "is Rosario Simone, a journalist specialised in the Middle East. For about 20 years, since he studied Arab language and literature at the Institute of Oriental Studies at Naples University, Simone has published reports of journeys. The main aim of his work is to bring the two shores of the Mediterranean closer culturally. This report is the fruit of a recent journey through oases of the desert of Mauritania, usually only in the news once a year during the famous Paris-Dakar rally. The oases guard an ancient treasure of the world cultural heritage."

They were places Rosaluna and her new friend Sinbad had been to during their cavalcades in northern Mauritania, part of a journey into the deep south of Morocco a few days earlier.

The loves of Rosaluna

Through tales told by her grandfather and family attitudes, Rosaluna developed a profound environmental consciousness. She had great curiosity for the sciences and geography and was attached to her Sienese origins. She was convinced that the great capitals of the world were the small provincial cities, provided that culture could supplant pettiness and opportunism and that the opportunities brought back from journeys could feed souls and minds. Though well aware of these things, as a young woman she needed new experiences, new teachers and the occasional transgression.

The trip to the caravanserai was in a way an entropic watershed between fine linear knowledge and far-from-equilibrium knowledge, or to say it with Prigogine, when equilibrium and order arise from chaos, often in a completely unpredictable manner. Rosaluna knew that she could sail the open sea in search of new destinations, but at her very young age, her great wisdom did not press her to emulate Ulysses in the conquest of new but unidentified horizons, but rather invited her to seek "fine and beautiful things," in the words of Archimedes, without losing sight of the lighthouse *of beauty*.

She had recently begun to read *The Novices of Sais* by Novalis, a master of the German Romanticism, and had been particularly struck by the words of the Master:

> *The voice that spoke was certainly that of our master for he knows how to link traces dispersed here and there. A certain light shone in his eyes when he showed us the ancient runic alphabet and he watched us closely to see if the light that makes signs clear also shone in us. If he saw us sad because the night did not leave us, he consoled us and promised future happiness to diligent tenacious observers.*

> *He often told us that when he was a child, he had a compulsion to use his senses, to busy them, to satisfy them. He observed the stars and traced their positions, their orbits, in the sand; he watched the sea of air, never tiring of considering its clarity, motion, clouds and lights. He collected stones, flowers, insects of all kinds and ordered them according to different criteria. He was a careful watcher of humans and animals, and looked for shells on the seashore. He eavesdropped his soul, his thoughts, determined to ignore where they would drive his desires. Once an adult, he began to travel: he visited other countries, other seas, he got to know climates, stars, plants, animals and men. He explored caves and observed the different coloured strata of earth. He moulded clay into strange forms. He found known things everywhere, though mixed and combined in extravagant ways, and thus the most singular things came to be arranged in a special order within him. Later he began to dicover links, relations, coincidences in everything. Soon he ceased to consider things in an isolated manner. In this open colourful framework, his capacity to perceive through the senses grew: he heard, saw, touched and thought at the same time. He enjoyed connecting distant things. The stars could be men, and men stars, stones animals, clauds plants; he played with forces and phenomena and knew where and how to find or evoke them. He knew how to strike sounds and chords from strings.*

These words were so far from the instrumental rationalism that they taught her at school and created fertile soil for two encounters, two great loves: Altair, the white horse with the golden star and the young Mauritanian Sinbad. Altair was not only a faithful friend for galloping on the beach or riding in the woods, he was a magic horse, descended from unicorns and able to go back in time.

Thus, Rosaluna's masters multiplied with journeys in time and included Novalis, Prigogine, fra' Benedetto, the prior of the convent of San Marco in Firenze after Savonarola, and the master of ecology of mind, Gregory Bateson. From the latter she learned that "conservation without evolution is death, but evolution without conservation is madness." Since then, Rosaluna called herself a conservative revolutionary, specifying, as Ernesto Cardenal had personally told her several months earlier, that the socialist revolution would fail without Christ and that peasant culture was deeper and more complex than working class culture.

Altair taught Rosaluna to value wonder at the colours and perfumes of nature. Once he took her among a herd of bison in the American prairies. Rosaluna met Tatanga Mani (Walking Bison). They sat on the grass and between puffs on his pipe, he reflected, "We see the work of the Great Spirit in almost everything: the sun, moon, trees, wind and mountains. Sometimes we appeal to it through these things. Nature is the book of this great power that could be called God, though we call it Great Spirit. What difference in a name?"

Rosaluna frequented the American Indians constantly with pleasure and benefit. She also established friendships with certain old friends of her grandfather, such as Birgil Kills Straight of the Lakota Nation, who always concluded his letters with *Mita Kuye Oyasin* (you are all my brothers), and Osceola of the Seminoles (*he who touches the enemy in battle without a bow*), the hypnotiser of alligators.

Altair had initiated Rosaluna into shamanism. The golden dunes enhanced the horse's white coat and Rosaluna's turquoise robes. A small dot on the horizon grew larger and Rosaluna distinguished a horse and rider. Rosaluna had learned to recognise Arab lineages: this was a Beni-Hassan prince of Mauritania. Like the Arab gentleman at the caravanserai in Fez, he wore a splendid aquamarine pendant. The young rider bowed to the girl.

"I am Sinbad of the Beni-Hassan," he said flashing his coal black eyes. By way of explanation, he added, "My mother is a Saharawi and I was raised on *The Thousand and One Nights*. She loved the actor Douglas Fairbanks in the film *Sinbad the Sailor*. The source of my name is therefore banal. If the horse you are riding is Altair, you must belong to the family of the great Magus of the three chests from Siena, friend of my princely uncle and a scientist respected by many Arab families."

"My grandfather," said Rosaluna, "Shall we go back?"

The sight of the two young riders on the dune crests with the sun behind them would have inspired even a non-rhetorical photographer such as Toni Thorimbert, a dear friend of Rosaluna and Giovanna. They rode to the estuary of the Rio de Oro and Levriero Bay. As they made for the nearest caravanserai, the two soon found much in common and had hours to talk. Sinbad spoke Italian well (he had a Corsican grandmother) and he told an interested Rosaluna about the *Capitana dello Yucatan* by Emilio Salgari, breaking of the enemy lines at the port of Santiago di Cuba and the struggle against the American empire, which was even then extending its influence into the Caribbean. They spoke of the beauty of the many Mediterranean ethnic groups and of the cultures that enriched them over the centuries. They spoke of talismans and sacred symbols, stopping to kiss in the shade of a *talha*, the gum arabic tree. In bidding her goodbye, Sinbad gave Rosaluna three gifts: a turquoise for adventure, a lapis lazuli for biodiversity and a moonstone for the love that was to flower in the years to come.

Aigues Mortes

Veliero Edoardo Verdiano had wanted to go to the Camargue in search of the old gypsies, but in those lands that devour stories and characters, the latter seemed to have disappeared into thin air, perhaps as happened to Mary Magdalen centuries earlier. He and Rosaluna stopped at Aigues Mortes, more as tourists than anything. The elderly Tuscan gentleman was interested in the town founded by St. Louis IX: the port where the Holy Women disembarked on their flight from Palestine; the port from which the seventh and eighth crusades embarked for the Holy Land. That day, in the market street of Aigues Mortes, there was a stall selling old wares.

"Bienvenida Rosaluna," said the man on the stall with a typical Spanish accent of Latin American, "I have something for you."

Rosaluna started. She did not know the man.

"You are Rosaluna, are you not?" he continued, "the girl with the tiger behind the left ear."

"Well, actually . . . ," Rosaluna replied, incredulous. The man handed her a small brass mirror decorated in bas-relief with a two-tailed mermaid. Rosaluna regarded herself. Behind the lobe of the left ear, a small tiger was clearly visible.

"Here!" said the man, offering her a Galilean kaleidoscope in white cardboard trimmed with brass. "And these are your Lari and Penati in dark green Guatemalan jade. Your friend Gamusa left them here last month, before returning to Yucatan." The man bowed to Rosaluna and added, "I am Orghelis. I make kaleidoscopes. The magic is yours!" Then he set off towards the walls of Aigues Mortes, disappearing into the mist.

Many years later, Rosaluna was involved in various adventures with her Creole-Garifuna friend, Gamusa, in the Yucatan Peninsula of Guatemala, and remembered Orghelis.

She returned to the Chianti rich with new experiences, with her splendid horse and some precious talismans. In the streets of Aigues Mortes, a girl had given her an old parchment inscribed in Latin. It described the magic square in the walls of Siena cathedral, a mysterious alchemical writing about Mary Magdalen, Siegbert IV, Bera III and Gaudi's flowery church in Barcelona. Rosaluna liked Gaudi, but did not like secrets. She crumpled the parchment into a ball and threw it down the first well she passed. In order to grow a little more, she probably needed the sacred Pacina garden, the perfume of lindens and her grandmother's tales.

A few days later the telephone rang. Sinbad had managed to tap into one of the telephone lines laid in the desert for the Paris-Dakar rally.

3
The land of birds

Rosaluna was raised on honey, pine nuts and stories. One evening her grandfather told her the fable of the Land of Birds, and pedantic as he was, before telling it he said, "This is a fable for children and a metaphor for adults. It is about pelicans and men and the ancient values of civilisations based on trade: Phoenician merchants who used shells from Cyprus as money and traded purple pigment throughout the Mediterranean, Venetian merchants, Marco Polo and the Etruscans before them. It is about the sacredness and biodiversity of nature and also contains a eulogy to that great French invention, the bidet."

Polybio the pelican

Once upon a time there was a pelican who lived in the Caribbean. He had an iron memory and his friends therefore called him Polybio. Asked his original name, he would always say that he could not remember. Whether this was a caprice or the truth, no one could ever tell. In any case, Polybio was the historical memory of the pelican nation.

In the evening, perched on his favourite branch of the tallest palm in Half Moon Cay, he would tell young pelicans about the adventures of ancient migrations and the encounter with the human nation. Only he could speak the language of humans.

He would light his amber-coloured briar pipe which he filled with his favourite tobacco. Polybio had a dear friend of the pelican nation of Cuba, who would bring him the best tobacco stolen from the plantations of Pinar del Rio, the tobacco of the most exclusive cigars in the world, Cohibas and Montecristos. His Cuban friend was known as Montecristo, perhaps after the name of the cigars or perhaps because like the Count of Montecristo, he tried to right the injustices of humans and pelicans.

One day when Montecristo was visiting Half Moon Cay, the two pelicans were exchanging the latest news circulating among the bird nation. According to this news, the climate was changing faster than ever before in bird memory. Hurricanes would increase in number and force. Confirmed by albatrosses and frigate birds, this news worried Polybio and Montecristo a lot. They knew that they lived in an area subject to hurricanes and cyclones.

"We shall have to organise a new migration of the bird nation like they did 10,000 years ago for the great glaciation," concluded Polybio.

As suggested by Montecristo, next day he called in on his friend Francesco, of the human nation, who had come from Europe for his annual visit to his Garifuna friend, Gamusa, who lived at Caulker Cay, Belize. Francesco was as prudent and scholarly as his friend was reckless and adventurous. Gamusa got by day to day. Besides fraternal friendship, they shared the conviction that it was not the laws of men that mattered, but native laws based on history and knowledge of nature. Gamusa claimed to feel his laws when at sea, listening to the wind and the songs of the manatees. Francesco, on the other hand, trusted his stormglass with camphor crystals.

Polybio landed in the giant ficus, the only tree of its type surviving the last cyclone. The 1962 hurricane had divided Caulker Cay into two. Francesco and Gamusa were smoking their cigars as they reclined in comfortable Rio Negro hammocks. Polybio greeted the friends and primed his pipe with tobacco.

The archipelago of 11 islands

After his discussion with Francesco and Gamusa at Caulker Cay, Polybio realised that there was no time to lose. Back at Half Moon Cay, he and Montecristo gathered any of their friends present on the island and nearby islands. Polybio told them everything he had learned from Francesco. The climate was changing and birds would have to migrate as they did 10,000 years earlier.

Polybio and Montecristo suggested that the problem be tackled in three phases: a first phase of wide-ranging exploration with the help of the albatross nation and the frigate bird nation, who had much greater flying ranges than pelicans. Once a place to migrate to had been found, the second phase would consist in gathering all representatives of the bird nation at high risk in their present sites. Whether fish-, fruit-, nectar- and insect-feeders could migrate there depended on the habitat and resources of the new place. Polybio proposed building a new land for birds with the help of friends living far from the land of pelicans.

The third and most difficult phase would be to build houses, shelters and nests in trees to defend themselves from climate change, bad weather and storms, ensuring a future for bird generations to come, a type of sustainable development that also considered a future for the bird nation. Polybio realised that without the help and scientific knowledge of Francesco, they would be unable to achieve such a great task. At the meeting, it was unanimously agreed that the three phases would last a year and that the third phase, including the building of tree houses, would in any case have to be completed before winter.

There was no time to lose. To begin the first phase they divided tasks. The gull was given the task of visiting different seaports to gather information on where there could be islands unaffected by hurricanes and cyclones, on the basis of climate maps that Francesco had given Polybio. It was necessary to spread the word and obtain information from out of the way places and fishing grounds, from birds who might know about new islands where the bird nation could live. They had to be out of the cyclone zone and off the normal shipping routes of the human nation. In the first three months, the gull had to report on stories told during long evenings in pubs, bars and ports, along seashores and on islands. Montecristo had the task of inviting the Great Albatross, historical friend of the pelican nation, to Half Moon Cay in three months' time. He would also invite the three frigate birds, Nina, Pinta and Santa Maria (with evident reference to the caravels of Columbus) who knew the Atlantic like the insides of their own pockets. The information that the Albatross could obtain from Patagonia and Antarctica, together with the information provided by the frigate birds on the Atlantic, would be useful in the first phase. Clearly, Polybio would not in the meantime rest with pipe in beak or wing in hand, but together with many young pelicans would speak to Francesco and Gamusa who would obtain information on the situation from the human nation. The young pelicans could also perch on the masts of ships and gleen information on shipping routes from the Caribbean to Africa and Europe. The next meeting was to be at Half Moon Cay in three months' time.

Anyone near Half Moon Cay three months from then would have seen a most unusual sight and ornithologists would have jumped in their skins. Birds of all sizes arrived from all over the world. Birds came with thousands of kilometres of travel behind them and others arrived from the mainland and nearby islands: northern birds, equatorial birds, Antarctic and tropical birds; among them, the majestic flight of several albatrosses escorting the Great Albatross who landed at the Cay, joyfully greeting his friend Polybio.

The first summaries were drawn and several hypotheses were excluded. The birds coordinated by Polybio, Montecristo and the Great Albatross listened to the detailed accounts of the frigate birds and the gull and gathered information from the young pelicans. This gave them a good picture of the situation. The most precious information, brought by Santa Maria, confirmed two strange rumours heard by the gull. The first came from a pub in Roundstone, on the Irish coast, where they serve the best crustaceans in Ireland with excellent beer. The rumour the gull heard in Ireland was in line with a second rumour originating from a port in Patagonia. The rumour hailed back to some Italian sailors who told of a group of islands related by other sailors. The Italian sailors had never seen these islands because they were off the conventional routes. The main thing was that the gull's information and that of the frigate birds agreed with other information obtained from wise old Timothy, a sea elephant seven metres long and as wide as a boat. It also coincided with stories told by turtles that return to lay eggs on the beach where they were born 20 years before. All these stories agreed with the tale of the Great Albatross that there exists a group of small islands, completely uninhabited and full of trees and plants, off the normal routes, and far from human battles. Reference was made to that ridiculous war for the islands known to Argentinians as Malvinas and to English as Falklands. The group of islands was even far from these, in a cool but substantially temperate climate, where extreme events, such as hurricanes, cyclones and tornados, could reasonably be excluded. They were therefore safe as far as climate and shipping routes were concerned.

In the days that followed, the birds compiled, checked and pinpointed latitude and longitude. Polybio filled many pipes and Montecristo drank much rum. There was also a discussion on the best brand of rum: Zacapa Centenario from Guatemala, Matusalem from Cuba or Barbancourt from Haiti. In the end they had exact confirmation of the coordinates of a group of 11 small islands, definitely covered in vegetation.

The next problem was to assess resources: the type of trees and berries, the presence of freshwater and fish. It was a miracle that these islands existed. While these aspects were being discussed, seven albatrosses, friends of the Great Albatross, arrived from Africa via Patagonia. They had been intercepted by some of Polybio and Montecristo's messengers. Hearing the news and knowing the places, they decided to go to Half Moon Cay. They arrived at a good time and the news they brought about the 11 islands was most auspicious. As the gull had heard and as the frigate birds had reported, they were beautiful islands in a sea teeming with fish, perhaps because they were off the main routes. The albatrosses had exact information about plants and the variety of species producing berries. In the middle of the largest island, there was a great natural basin in pink porphyry, hidden among the trees. Rainfall was good and the basin was full of freshwater throughout the year. The Great Albatross explained that there was a system of siphons by which excess water went into cool underground chambers, replenishing the basin when the pressure above decreased. The sea abounded with crabs, crustaceans, molluscs and fish of all kinds. There were many insects on the islands which seemed a paradise for birds in general. Polybio was happy to know that there were many flowers on the islands, including many species of orchids and especially the epidendrum called *bella di notte* because it emanates perfume at night. These were excellent for hummingbirds. Finally, the youngest albatross reported that some wonderful marine mammals, known as manatees, reproduced around the islands and guarded them as a sacred place.

The research and exploration phase was at an end. The place to migrate to had been found. The birds could proceed to the second phase. That night Polybio and Montecristo slept peacefully, knowing that the second phase would begin the next day.

A conversation on a boat and eulogy to the bidet

Francesco and Polybio were sitting on the bridge of Gamusa's boat, smoking a cigar and a pipe, respectively. Francesco told Polybio the story of Herman Daly's boat that was loaded with an optimal distribution of goods. Careful stowing enables the load to be maximised, but during the operations, nobody noticed that the boat was sinking because its carrying capacity had been exceeded.

"You see," added Francesco, "the question is tricky, but the underlying principle is important: one must not consume too many resources or produce too many wastes. A country, an island, or the planet Earth is like Daly's boat. Infinite growth is not possible on a finite planet. Because the oil magnates have upset the climate with greenhouse gases, you have to migrate."

The two remained a while in silence, contemplating the clouds. Francesco blew two or three smoke rings and said, "When you build houses and nests, keep this principle of sustainability in mind and always leave some wild land. Humans, too, can build beautiful sustainable houses or hideous concrete boxes. To tell how civilised a people is, you just have to see how they build their bathrooms. Civilised people use the bidet, a great French invention, with a central tap that mixes hot and cold water. Barbarians have two separate taps and no bidet. Barbarians waste water and energy but civilised people build their houses with care, making optimum use of solar energy and observing the principles of bioarchitecture."

"Take it easy, Francesco," the booming voice of Gamusa interrupted this refined conversation.

The hurricane house of Belisario Martinez

Three days after reaching the new islands, an assembly was held to discuss the construction of houses and nests. In those three days the birds had got to know the islands and the weather had been beautiful. It was necessary to build nests before winter. Polybio was recognised as the sage who had conceived the idea of the great migration. Three other members of the bird nation sat beside Polybio. The Great Albatross would not live on the islands but was willing to offer his wisdom and knowledge. Montecristo was the second and the third was a beautiful, intelligent, astute hummingbird from the forest of Monteverde, Costa Rica, and represented the land feeding birds. The assembly decided to elect a government of nine components that would hold office for a year and could not be re-elected. The four sages would not be part of the group of nine, though respected and heeded by all members of the bird nation. The nine would be birds with qualities that history recognised in Julius Caesar and the emperor Hadrian, and in that happy bird nation there certainly were no stupid demagogues like Brutus, or envious smart alecs like Cicero. The four sages were respected because they did not manage money or exercise power but were simply birds who knew many things which they handed down to future generations.

The assembly decided some very important things. Four of the 11 islands would not be used, at least for several generations. They would remain wild so that their biodiversity could continue to reproduce in a completely natural manner, to ensure the multiplication of insects, berries and rare fish and plants that were necessary for life. The only action permitted on these islands would be visits to bring back pollen or seeds to enrich the seven islands where the houses and nests would be built. These seven islands were assigned on the basis of size and number of birds, but mostly in relation to their characteristics. The one with most flowers would be home to the hummingbirds, who could of course visit flowers on other islands as well. The islands with the most fish would be inhabited by pelicans and other fish feeders.

The time had come to build houses. Polybio addressed the young birds, telling them that the migration had been successful because they had been prudent. He praised the wisdom of his friend Francesco, without whom they would not have been where they were at that time.

"We exploited a sentiment very important for knowledge, namely fear," continued Polybio. "We made this journey convinced of the need to escape from a situation that humans created, which has changed the climate of the planet. We acted out of fear of extreme events that will soon strike the areas we used to inhabit. In this way we took the path of lesser evil. The mighty Hannibal counselled Scipio before the battle of Zama, but Scipio did not listen. Hannibal said it was always better to choose the lesser evil. And so we are here. Thanks to what we have done, future generations will be able to enjoy a life of joy and quality."

Of the youngest birds, Polybio asked, "How many of you have dreamed of having a wooden house in the trees or a safe refuge among the vegetation? Each of us will build a house. Fishers will build houses where they can dry fish in special storage spaces. Others will build nests. The idea is to use vegetation resources to a maximum, building houses and nests close together among the vegetation so that they afford protection from great storms. If we use our ingenuity, we can weave villages from the bottom up using twigs, mud and earth. This will give us comfortable houses and storage spaces for food, and make sure that fledglings are not blown away by the wind. We should begin from inside, where the vegetation is thickest, and build from the bottom up. These houses will be protected by external layers if we work in concentric circles. This type of structure will also be used on the other islands. Everyone will build his own house as he wishes but there will be a golden rule to follow. Besides being functional and offering protection against storms, the house must also be beautiful. We will also drop seeds of colourful flowers to beautify our surroundings."

"You must not forget that we are bestowed with a treasure, the natural porphyry basin that provides us with fresh water all year round. The basin must be conserved and never dirtied. The houses must be sufficiently distant so that the basin remains the sacred place of water, where we can bathe and drink, but not damage in any way. It is a magic place and must never be contaminated with the wastes of our houses.

Let fear of storms guide you. Here there will never be cyclones or hurricanes, but there will certainly be storms and heavy rain in winter. In the next three months we have much work to do. We must work and obtain help from the albatrosses who can bring sand, stones and wood. We will work together on the seven islands. We will also be helped by our manatee friends, who can bring sand to make mortar for our nests. This will be useful for the swallows who build their nests under the roofs of the wooden houses of the fishing birds. We will ensure that food is replenished for future generations, multiplying crops by fertilising the soil with our excrement and sowing new seeds to vary the quality. We will bring pollen from one island to another to cross-fertilise flowers.

Our 11 islands will be open to friends, not only albatrosses and frigate birds, who will be leaving us in a few months, but also to friends of the human nation, for whom we will build houses because guests are as sacred as water. We will let these humans sojourn on our islands, living according to the laws of nature, so that the laws of the 11 islands, which today are the laws of nature and not artificial laws, can be the laws that Novalis referred to in *The Novices of Sais:* the laws read in the stars at night, in the shells and sand during the day and in the forms of butterfly wings and the colours and biodiversity of flowers. We have migrated, but not like Ulysses, who recklessly went looking for adventure beyond the columns of Hercules. We were guided to take a necessary step by the wisdom of Francesco and our sages. With prudence and wisdom, we fled from danger, choosing the lesser evil. Adventure is fine when it does not end in death. No bird ever dared to make such a long journey, and it would not have been possible without the knowledge of Francesco and Gamusa. With the help of these humans, who know the sea and who are friends of the bird nation, we chose this archipelago, off the shipping routes, because we want nothing to do with unsavoury characters.

We will host friends of the human nation and trade with them. These islands are also rich in things that do not interest us, but that humans may wish to trade, like the merchants of Venice, the Phoenicians and the Etruscans. The Phoenicians obtained purple dye from the *murex* shell and traded it in the Mediterranean area. The Etruscans cooked earth in special ovens and made beautiful black earthenware known as *buccheri*. The Venetians exchanged goods and knowledge with Cathay, thanks to Marco Polo.

There is fraternity between our nations, a community of living beings, and some humans have been thinking of community with future generations. Indeed, one reason we undertook this great journey was because we do in fact consider future generations. This is something we must never lose."

Montecristo then came forward. "Fledglings, I want to use my turn to tell you a story. Once there was a friend of the pelican nation who belonged to the human nation. His name was Belisario Martinez and he lived on an island, Caulker Cay, near where we used to be. He was one of the first humans to live there, together with a few other lobster catchers and the first merchants. Belisario knew the art of catching and cooking lobsters. He served them with a special sauce containing garlic, black pepper and a strange hot spice mixture that his mother taught him. His mother was a strange woman. Of Chinese father and Portuguese mother, she came to the Caribbean from the Portuguese colony of Macau. It is also said that she was granddaughter of the famous Yanez de Gomera. Belisario's father was a mestizo, part gypsy, part Black Carib (Garifuna), who were the first cross between black slaves and Amerindians, the people of the islands that Columbus discovered on his first voyage."

Stormglass

"With the money he inherited," continued Montecristo, "Belisario bought land near the sea on the ocean side, in view of the coral reef where the surf crashes. The land had many coconut palms and vegetation right down to the white sand of the beach. The house had two storeys and was built on stilts, as is customary in those islands. It had a wooden verandah where Martinez spent a lot of time. One afternoon, Belisario was contemplating the sea, which was completely calm. There was not a breath of wind. Some clouds were drifting towards Belize from Mexico. Belisario noted the turquoise sea, the bright green of the palms, and many birds in the sky. Suddenly an emerald hummingbird was vibrating in the air a few metres from the verandah. Coloured birds chased each other and sang among the palms. Then his eye was caught by several hundred gulls gathering near the wooden jetty where boats moored. He got up and went into the house to get his antique stormglass, a glass tube containing camphor crystals, the best instrument for predicting storms. It records changes in electromagnetic forces many hours before a storm arrives. These forces make the camphor crystallise inside the oily liquid that fills the tube. The instrument was invented by Fitzroy, captain of the *Beagle,* the ship on which Darwin sailed around the world, realised the importance of biodiversity and formulated his theory of biological evolution.

Belisario Martinez noticed that the crystals were forming faster than he had never seen before. A hurricane was surely coming. It was then that Belisario decided to build another house for future hurricanes. He built it inland in a sheltered spot, just as we will have to build ours on our planet of 11 islands. He built a very low house with strong poles of very hard wood, driven deeper into the sand than the roots of palms. He used his head and his muscles. Belisario was young then. He called it the hurricane house. He would go there whenever a hurricane struck Caulker Cay.

Belisario had a parrot who perched on his shoulder and came with him whenever there were hurricanes. The hurricane house was open to all humans and animals who wanted to be safe. Belisario was very wise and when he grew old he passed on many secrets to a friend who helped us get here: Gamusa. Gamusa was a wild young Garifuna boy, whereas Belisario was old and wise, heeded and venerated by all. The many wrinkles on Belisario's face indicated his many adventures. He was an attentive observer of nature which was one source of his wisdom. He loved the island where he grew up. Because his family had no place to live, he wanted them to find a homeland and this was one reason he built the hurricane house. The story of Belisario Martinez is a story of many women and many children. Some of them stayed on the island and other wandered the world, following the call of some ancestor. The story is not finished because Martinez will live to be 120. By then, he will have too many wrinkles to count.

The reason I wanted to tell you this story is because we have to use some of Belisario's wisdom on our islands and build many hurricane houses, even if hurricanes like that should never reach us and the forces of nature should be on our side. However, the wisdom of Belisario will bring us luck in founding our 11-island community."

The young birds listened to Montecristo's story and their imagination set to work. Some dreamed of Chinese junks sailing east, others of a parrot to carry on their shoulder, others of Belisario's beautiful women. Some imagined the friendship with Gamusa, others the strength of the hurricane house, the only house never destroyed by a hurricane in the long history of the Caribbean. Just then, four pelicans arrived with good news for Montecristo and Polybio. They came from the second auxiliary boat, that of Gamusa. The news was that nobody had noticed the route taken by the two boats and nobody had seen the great migration of birds. Ornithologists had remarked a decrease in certain species in certain islands, but Polybio's strategy of having various species come from different parts of the world, so as not to upset previous equilibria, had been successful. The ornithologists had noticed nothing. Francesco and Gamusa sent their regards to Polybio and added, "Suerte! For the archipelago of 11 islands."

Rosaluna had fallen asleep. Her grandfather whispered that next time he would tell her about the rum punch of Belisario Martinez, about pearl merchants, and about certain gentlemen of fortune who were Gamusa's friends.

4

The alchemist of Mt. Amiata

Colours and Entropy

Nature sometimes
to alchemy turns
inventing colours
full of magic.

It mixes manganese
with the mountains
to make stones
a kindly faded pink
like trachyte
in the Bosa caves.

By salts of nickel
and mines of cobalt
it turns the grey basalt
into azure.

When the fog comes
it chooses rare earths
and with indigo paints
the mantle of the sea.

It uses of the elements
the rare metals
and before the merops

comes to rest,
behind her back,
it paints a dress
of silver and gold
woven centuries ago.

It uses red of iron
for blood and life
but if metals
from its fingers slip
entropy mixes them
and makes them white.

This always happens
when Nature gets tired.

A scientific experiment in Via Mezzocannone

At least three versions of the first scientific experiment exist. Conducted in Via Mezzocannone in Naples, it led to a Nobel Prize in Medicine in 2018, about 30 years later. It all began in 1520, when fisherman Giorgio di Iorio fell in love with the two-tailed mermaid. In those days, strange things occurred in the sea between the islands of Ischia and Procida. It seemed that many prophecies of Hermes Trismegistus and the Sibyl of Cuma had taken flight from the inlaid stone floor of Siena cathedral and gone to roost near those islands, not far from the sibyl's grotto.

In those days, only a few families of poor fishermen and grapes or orange growers lived at Sant'Angelo d'Ischia. Giorgio di Iorio was young and restless and used to sing and talk to the sea. This habit of his was to change his life.

A number of rare coincidences happened at the same time. The scirocco was howling and warming the waves, Giorgio hauled up a small seahorse and let it go, a rainbow touched the sea in that very place and a group of dolphins began to swim anticlockwise against the vortex of the scirocco. Giorgio was afraid and spoke to the sea. The water replied to a human for the first time. Its voice, at first indistinct, then increasingly clear, a female voice, reassured the young fisherman and spoke of things that Giorgio could not understand: two liquid structure domains had formed under the influence of different effects, one natural (the dolphins) and one quantistic. According to the voice, the second did not produce entropy. Giorgio was terrified and began to sing a Neapolitan song at the top of his voice.

It was 1520, and for the first time, a mermaid flipped out of the sea onto Giorgio's boat, enchanted by his voice. She had everything in place, beautiful breasts, turquoise nipples, velvet skin and eyes the colour of the sea. They made love for 11 days and 11 nights, then the mermaid swam down into the inky depths followed by the seahorse and Giorgio never told anyone about what had happened.

The first version of the experiment conducted in 1988 in Via Mezzocannone was reported by Professor Vittorio Elia of Federico II University in Naples. According to the theory, water had a sort of memory that remained during thermodynamic and calorimetric experiments. People who used the equipment left signs of their presence.

The second version, conducted by physicist Emilio Del Giudice, was concerned with quantum coherence domains. The third, conducted by Rosaluna's grandfather, was concerned with Prigogine dissipative structures and their self-organising capacity. The three versions actually had something in common: the supramolecular structure of water, its fluctuations in the presence of living beings and the protagonism of this small molecule in biological affairs. If this is true, the three scientists said, we can begin to understand why homeopathy works even in the absence of a single molecule. The active principle leaves its sign in the supramolecular structure of water.

Thirty years later, when the doctor from Siena, dressed in his very best, was about to be awarded the coveted prize, he could not get a childhood phrase out of his head: "Once upon a time, there was a king!" Once upon a time, there was a king. Was it fable or was it real? He pinched his cheeks. He could not tap his pipe on the ground in that situation. King Gustav of Sweden was standing before him to award him the Nobel Prize.

It should be recognised that some merit for those discoveries was also due to an alchemist of Viareggio, or rather an affable *spagirista*, who concocted health-restoring quintessences. The whole story cannot be told here, but if someone were of a mind to write it, they could call it *The Newborn Palermo Soccer Club.* The Nobel laureate of Siena and protagonist of these events tells it thus:

> *One morning, 45 years ago, I was keeping a general surgery patient anaesthetised, when I was called next door to the delivery room to resuscitate an asphyctic newborn. Despite the usual first manoeuvres by the midwife, the baby did not begin to breathe. He was cyanotic and flaccid. I intubated him and administered oxygen, checking ventilation, and soon observed an extraordinary phenomenon. Exactly one longitudinal half of the baby had become pink, whereas the other half remained grey and cyanotic. Continuing the oxygen, it was not for another 15–20 seconds that the pinkness spread to the whole body. Then the baby began to breathe spontaneously. I aspirated the mucus, removed the tube and the baby began to cry as they all do at delivery. In every other aspect, he was completely normal.*
>
> *There were no problems in the next few days. The pediatrician excluded neurological abnormalities and malformations. Only the episode at birth was abnormal, unexplained: a two-coloured baby, with one side pink and the other black, like the jerseys of the Palermo Soccer Club. I have often thought about this incident and it took me a long time to reach the only possible interpretation. I studied what little literature there was at the time on the dominance of the left and right hemispheres of the brain, on apparently unrelated fields (energy flows and Chinese medicine, biorhythms, neurotransmitters, homeopathy, astrology, etc.) and later tested and developed some diagnostic and therapeutic methods that I still use.*
>
> *The common denominator of such diverse fields lies in three concepts that are closely interwoven like the double helix and bridges of DNA. The first is that the human being is a microcosm potentially in harmony with the macrocosm that surrounds him. Second, our lives go normally if the biorhythms and energy flows within us are normal. Third, diseases, apart from genetic ones and external aggression, stem from blockages, deviations and weakening of our vital energy. The big picture must always be considered.*
>
> *Despite great technological progress, contemporary medicine unfortunately reduces a sick human to a metritic uterus, an ulcerating duodenum, an arrhythmic heart . . . . Most treatments suppress symptoms, while university teaching and medical practice do not take*

a holistic view of humans, either healthy or ill. Medicine has gone backwards in this sense because up until 50 years ago many teachers of my generation bore it in mind and tried to teach it, bearing in mind the importance of general culture for their students.

To return to the asphyctic baby, it is evident that the situation of the two halves, paternal and maternal, in each of us, remains constant during embryonic, fetal and adult development. Radioesthesia shows this. The pendulum swings to and fro when brought beside the temporoparietal region of the head, continuing over the trunk on the same side and the external dorsal parts of the limbs. It swings in a circular pattern when brought near the same sites on the opposite side. The opposite occurs on the ventral surfaces of the limbs and posterior trunk. To and fro and circular patterns of the pendulum express two different energy polarities: sex-related differences; the analogy with the Chinese concepts of yin and yang is immediate. The dominant sexual hemisphere is generally the right in women and the left in men.

I have no news of that baby. I think it likely that he would have been susceptible to chronic inflammatory disorders of organs on his grey side rather than on his pink side.

This was the scientific atmosphere in which Rosaluna grew up at the turn of the millennium among her grandfather's Sienese friends. She knew she had two alchemist ancestors. Fra' Benedetto was elected prior of the Convent of San Marco in Florence after Girolamo Savonarola, and like his predecessor, struggled against powerful Medici popes. In 1529, the monk mounted the glacis at the siege of Florence, inciting souls with his words and fiery eyes. Before being taken prisoner in Castel Sant'Angelo, he left an unpublished manuscript on "spagiria" and alchemy which came down to Rosaluna's pharmacist great-great-grandfather. She used to read the secrets and the family tales on the eleventh branch of the old magnolia. She would take a jug in the form of a hare, found in a local Etruscan tomb and used for a cover of the magazine "Arancia Blu." This ceramic jug was a work of art. Rosaluna would fill it with mulled wine containing a touch of cinnamon.

The other alchemist was Giovan Battista Becchini, known as Tista, Rosaluna's great-great-great-grandfather, once pharmacist at Arcidosso on Mt. Amiata and at Orbetello (Mt. Argentario). Rosaluna had relatively recent memories of him from tales written by her great-grandmother towards the end of the previous century. These manuscripts, which Rosaluna would only read in the tree, are published here, in their original form, in the style of the great grandmother. The first is about the pharmacist of Mt. Amiata and the other completes a picture of the period.[1]

The pharmacy of Arcidosso and St. Davide

My maternal grandfather was a pharmacist. Although he graduated in pharmacy and had a pharmacy in Arcidosso on Mt. Amiata, he was unlike his colleagues. Pharmacists are usually graduate merchants who wish to sell their products and try to convince their clients to buy certain concoctions for certain ailments. My grandfather, however, did not have much faith in the things he kept in his shop. He prescribed infusions of chamomile (even if not harvested the night of San Giovanni) for insomnia and poultices of linseed for bronchitis. He knew and loved hundreds of medicinal plants and kept a good supply at home, ready to treat members of his large family.

When I was small, I used to spend several months of the year in Arcidosso with my grandfather and aunts. I was friendly with our laundry woman, Evangelista. Arcidosso was famous for unusual names. There was a girl called Folla Plaudente (applauding crowd), a boy called Anelito (panted or yearned for) and a man known as Beppe Fottilo (Beppe fuck him). The latter was officially Giuseppe, but the nickname had a story. When he was about to be born, the future father, who already had six children, was sitting outside the house in waiting with his head in his hands. A merry woman came out of the house crying, "Cheer up, Tono, it's a fine boy! What shall we call him?"

---

[1] The next two stories were written by Lidia Pulselli Tiezzi (Rapolano Terme, Siena 1905; Siena 1996), the author's mother. See also: E. Tiezzi, *The End of Time*, WIT Press, Southampton (UK), 2003; E. Tiezzi, *The Essence of Time*, WIT Press, Southampton (UK), 2003.

"Call him Beppe, fuck him!" replied his father, overburdened by too many offspring.

Our laundry woman told the aunts about a poor relative with an ailing husband who went to the pharmacy with a prescription for a tonic for her spouse. My grandfather read the prescription and said, "Listen, Marietta, with the money you would spend on the medicine, go and buy some good meat for your husband and then go to Marianna's, and tell her I said to give you a flask of wine." Marianna was our factotum who somehow also found the time to serve wine in a shop my grandfather made available to her.

There was great poverty on the mountain in those days and poverty led to apathy. I remember going several times with an aunt to the house of a poor sick woman. I always tripped on the doorstep which had stones out of place. The woman's husband was an unemployed mason, but in his idleness could not bring himself to repair his own threshold.

Grandfather would say people did not need medicine but good food and a helping hand. And a dentist! Many women had bad teeth or no teeth at all. A popular verse went:

O montanara, chi t'ha mangiato i denti?
L'acqua gelata e i saccioli bollenti.

(Oh mountain girl, what happened to your teeth?
Cold water and hot chestnuts.

Boiled chestnuts were the staple food.)

Aristotle (not the philosopher but Marianna's husband) told of two friends who stopped in a tavern to drink a glass of wine. One asks, "If you were rich, what would you eat?" "A big plate of beans with pork rind! And you?" "I dunno . . . the same I guess." According to Aristotle, there was little money and even fewer ideas.

When he was older, grandfather sold the pharmacy and bought Laticastelli, a farm near Rapolano. Here his life began. The manor house, surrounded by various farm buildings and peasants' houses, was situated on a rise. An avenue of cypresses wound up to it from the main road. The view included fertile fields sloping down to the

Ombrone river. In the other direction, rolling grey clay hills gave way to dense woods, where pigs ran almost wild. The clay hills produced wheat and the woods produced much that was used by the peasants. Several families of peasants lived in charming farm houses with curious names: Spiritello, Spiritellino and Spiritellone.

I don't think my grandfather knew anything about agriculture, but he would always go round the fields, talking to the peasants and listening to what they said. He couldn't resist citing the odd verse of Virgil's Georgics or Bucolics, but the peasants understood, specially a certain Maggini who did not know Virgil but loved bees, which he kept with great care and spoke about with admiration and satisfaction, as if they were his dearest friends.

Grandfather would rise early, drink a glass of milk or get his dear faithful maid Beppa to make him a bruschetta with new olive oil. He considered this oil the prince of condiments. Indeed, on his letterhead one could read: Laticastelli Farm, prop. Dr. Giovanni Battista Becchini, producer of olive oil "better and cheaper than butter." After breakfast he would go out into the fields and country pathways, collecting interesting plants and minerals. Very little escaped his gaze. Wandering for hours by himself, he never felt alone but was one with nature. He conversed with all creatures, animate and inanimate; he understood them and felt understood.

Sometimes he scrambled over the clay hills. In places the ground was bare of vegetation, and fossil shells were exposed here and there. With a special knife like a surgeon's scalpel he would gently free large scallop shells, shaped like the square of Siena, and harmoniously whorled murices from their grey clay prison. These beautiful ancient creatures spoke to him of remote eras. Grandfather had strong philosophical inclinations; he knew the world and its ugliness; he compared ancient and modern civilisation, noting that there had been no true progress. This disturbed him but he knew how to be patient in adversity and how to smile; he would see humour in the simplest things.

There was a large cellar with rows of enormous vats at Laticastelli. Each peasant put his grapes in a particular vat marked with a big sign: Oreste, Benedetto, Lorenzo . . . . Grandfather put signs on his own vats too: Eleonora Duse, Lyda Borelli . . . .

Many months of the year were spent in idyllic peace, but Grandfather did not forget his properties at Arcidosso. When the summer heat came, the aunts left for the mountain and he would join them shortly afterwards. In the town he was welcomed warmly; many families were related to him in some way and old friends and acquaintances were many. The house was an enormous old building which could accommodate many families; we spent happy days there among affectionate aunts and amusing uncles. Grandfather's properties included a series of fields and irrigated vegetable plots. Around Arcidosso there were many brooks and springs. When I awoke in the mornings, the silence was unbroken except for the silver sounds of water. I would run to the window and open it to hear them more clearly, sounds that evoked cool glades. Nearest of all I could hear the gurgling of the Fonte della Vecchia, a fountain of wrought iron with many spouts, in the little square near the house.

Grandfather also owned a farm on Monte Labbro, which was another world: not a trickle of water, not a tree. Under the hot sun the earth produced short fragrant grass for cattle. At the summit were the ruins of the house of Saint Davide. On Mt. Amiata, Davide Lazzaretti was "Saint Davide." For a long time after his death he was remembered and secretly loved. He must have been a person of great magnetism. His rather confused doctrines of a sort of utopian mystic communism could hardly have had much power of persuasion, but people loved him and were fascinated by him.

In the house of one of our peasants there was a mysterious chest that was never opened. One day my girlish curiosity overcame the reluctance of the wife and she showed me clothes that the elders of the family had worn on that fateful day, 18 August 1878, when the followers of Saint Davide left Monte Labbro in procession for an unknown destination. When they reached Arcidosso they were ordered to halt; a rain of stones met them from one quarter and shots from another; Saint Davide was mortally wounded. Visibly moved, the old peasant woman showed me the red and blue cloaks, the pink and white veils, the coronets of flowers.

Filled with curiosity and awe, we would wander among the ruins of the rudimentary houses of the followers of Saint Davide. Here and there we came across the symbol of Davide, a Christian cross with an outward-facing crescent moon at each end of the horizontal, signifying that one way or another, everybody had his cross. Grandfather was reluctant to answer my questions on Lazzaretti and would merely shake his head. I think he felt compassion for the tragic fate of the Prophet of Amiata and sympathy for his utopian ideas.

I knew an old follower of Saint Davide, by the name of Tommencioni. Often when my aunts and I were climbing among the shady chestnuts, we would cross his path. He would be coming down the steep track with his mountaineer's gait. He was old, lined and angular but tall and erect; his lively dark eyes were full of sparkle. "Do you recognise us?" Aunt Iole would ask. "Yes," he would reply, "you are the Tiste sisters." Since we were of the family of Signor Giovanni Battista, they called us the Tiste sisters. Tommencioni was a man of few words, but when he spoke it was in verse. For each of us he would improvise a rhyme as a sort of compliment and greeting. What tranquil days passed in peace with nature! An hour passed with a book or a toy in the shade of the chestnuts was a daily pleasure; to climb up to the rock of the Vettoraia was more strenuous; to reach the Contessa's Field in search of raspberries was a walk lasting several hours. We all went on foot, of course; there were no roads and no cars. Anyone who wanted to reach the cross at the summit of Amiata had to go by mule. Late at night at the "Tonda" (rotunda) the participants would gather with the mules and muleteers whose job it was to lead the way and to get the mules, famous for their sense of independence, to collaborate. The mules would not wear saddles; on their backs, light but unwieldy pack-saddles would be arranged. In order to stay there one leg had to go east and the other west. At the end of the journey, one's legs were terribly stiff. The summit was attained at dawn as the view unfolded. Three provinces lay below, stretching away to the sea. Now it can be reached easily by car, but Amiata is not the same; it is now a network of roads, roaring motors, bars and restaurants, and there are empty cans everywhere. Grandfather would have been upset to see nature defiled in this way. Luckily he was able to enjoy a clean and simple countryside for many years. He died quietly one night at the age of 83 years, like a lamp out of oil.

The story of Merope

The five Becchini sisters had a cousin, Merope, the daughter of one of their father's sisters, aunt Maria, and a cousin of his, uncle Raffaello. It was a marriage between relatives and a happy one. Besides being a rich landowner, Raffaello was one of the best lawyers in Grosseto and she was a plain, plump, respectable woman and excellent housekeeper. Merope, the only child of parents who were getting on in years, was tended like a flower in a greenhouse. Local schools and homemade clothes were not good enough for her, only specially chosen teachers and excellent dressmakers: piano, painting, French and embroidery, everything necessary for young ladies of the time.

Merope was not beautiful but she was so well dressed and self-confident as to be striking. Were she simpler, she would have been nicer; her feeling of self-importance was reflected by an attitude of superiority and condescendence towards the poor common people around her. She was fond of her cousins, however, and they returned her affection and did not envy her.

The family lived in Grosseto but passed the summer at Arcidosso, and in those days, Merope was always in love. Naturally, at 18 there was already an aspiring fiancé, a young relative who was a promising lawyer, conscientious, religious and proper. Merope would have nothing to do with him: he was too serious, controlled and boring. She dreamed of romantic love and overwhelming passion.

One day, a young man came to Grosseto under unclear circumstances. He seemed to live from day to day and it was rumoured that his two sisters were skilled at acquiring the favours of various personalities. Young Giuseppe was tall and well groomed. He walked with a detached, gentlemanly air and cut a fine figure. He began taking his strolls in front of Merope's house and casting well-dosed glances up at a certain window. The young girl's heart took the cue and opened to this sudden, disinterested love, so romantic, based only on glances exchanged between the window and the street. Next came ardent notes exchanged through the agency of a complaisant maid. Passion blazed. Giuseppe stepped forward and asked the austere father for the girl's hand.

The parents were furious. A nameless individual with sisters who were the subject of gossip, a genuine hunter of dowries, had dared to cast his gaze on the rich, educated Meropina of good family, the most marriageable young lady in town.

Merope revealed her strong and decisive character: "I love him and want to marry him!"

Her mother was full of consternation. "How can you say you love him when you hardly know him? Perhaps you exchanged three words with him after the theatre the other night!"

"What would you know with your marriage of convenience! I, instead, know what love really is!"

A period of tremendous fighting and arguments ensued. Raffaello was an authoritarian and self-respecting man. His wife, servants, employees and most colleagues did not dare express their ideas to him. Raffaello adored his daughter but considered her a personal possession to place on a pedestal, of course the pedestal that he chose. For the first time in his life he became aware of his heart, and though it was hard and selfish, his heart bled.

He would stay back late at work every day in his office on the ground floor. Merope, on the first floor, would sit at the piano and play *The Prayer of a Virgin* and *Forbidden Music*. Raffaello listened, full of pride.

Merope locked the piano, threw out her books and embroideries and refused to leave the house. Though she had to sit at the family table, she refused to eat. She lost weight and her health deteriorated. What could the elderly father do?

"If you insist on ruining yourself, go ahead! Marry that ill-famed adventurer!"

The sun returned to Merope's face. The fiancé, now part of the household, was loving and devoted. A rich and opulent dowry was prepared.

In the meantime, cousin Elena had married and left Arcidosso. She often received letters from Merope, hymns of joy for the happiness won and of satisfaction at getting her own way. One day, a strange letter arrived: "Here in Grosseto there is an epidemic of measles. Even older people are catching it and falling dangerously ill. Let's hope I don't get it."

A few days later the news arrived that Merope was in bed with measles.

"But it can't be so serious! She will not be without medical care."

Indeed, she had plenty of medical care and after about ten days, her fever abated suddenly, perhaps too suddenly. Merope was weak and stayed in bed as suggested by the doctor. Her mother and a maid were in the next room, embroidering and whispering to each other so as not to disturb the convalescent, ready to attend should she need them.

Suddenly Merope awoke and felt breathless. "Why don't you come and help me!" she screamed. The two women precipitated into her room. Merope was standing erect on the bed like a ghost in her long nightgown, her eyes wide in her grey face, her arms extended. Suddenly the slender figure collapsed onto the bed in a flurry of lace and black locks. All was still and silent.

Now there is a pretty chapel beside the Convento dei Cappuccini near Arcidosso. A fresco shows an angel guiding a white clad girl to heaven. The girl's face bears a likeness to that of Merope. The votive garland beside the fresco carries the inscription "To a girl tenderly loved. Giuseppe."

Rosaluna read her great-grandmother's stories and imagined Merope's love and the acts of St. Davide. In the meantime, her alchemical knowledge of the biodiversity of water grew and she was increasingly attracted by the travel stories and adventures of uncle Vieri.

5

Gamusa

Suddendly the Pile-Built House

Suddendly from the sea rises
the pile-built house
like Montecristo from the autumn fogs,
but this is a pirate sea
among the Keys of Belize and Terranova,
Caia Luna and the great blue hole
of Cousteau and the stalactites.
The schooner approaches and you can get a glimpse
of a strip of land under the pile-built houses
and of mangroves that are dwellings for the Liguus
more than spiral shells of many-coloured pastels.
It's an island that appears and disappears
following the waves humour
with an only palm-tree rich in water cocoes.

There Gamusa lives,
a master mariner of few boats,
son of pirates and escaped slaves.

He speaks "garifuna" Creole,
mixing the Amerindo idioms
with ancient African tongues
and by a strong gesture of the hand
he tosses up the bowels
of the fish on which fling,

with a kite movement, two nimble terns.
He cooks lobsters just taken from their holes
and talks about long voyages
and divided half-moons
and sea nights.
The sand stripe
you can and can't see
gives hospitality to jealous liuto tortoises' eggs
laid during the last night.
A trunk in the bank mimes
an armadillo form.
Gamusa dreams of far children.

"Don't-Understand" peninsula

Yucatan lies between the Caribbean Sea and the Gulf of Mexico and includes a part of Mexico, Guatemala and what was once British Honduras, now Belize. Many ruins of ancient Maya cities, major archaeological wonders of the world, are found on the Yucatan Peninsula.

When the Spaniards conquered Yucatan they asked the Maya of Tulum and Uxmal the name of their tribes and cities and the reply was always the same: Yucatan. The Spaniards thought this was a place name, but the Maya were actually saying they did not understand: "Yucatan," I don't understand.[1]

The notes that uncle Vieri wrote many years after his two periods, roughly in the thirties of last century, in Central America, especially Yucatan and Costa Rica, had a beguiling title, *Lost World Revisited,* evidently after Huxley's *Brave New World Revisited.* Huxley frequented Siena for many years and wrote a small volume on the Palio. Vieri would certainly have met him at the Caffè Greco or in Piazza del Campo. The title referred to part of the archaeological park of Tikal in Guatemala known as *Mundo Perdido* and to the pyramid with four flights of stairs, one for each point of the compass. The most beautiful faces the pole star, *Xaman-Ek* in the Maya language.

[1]A. Melis, personal communication.

Vieri wrote of Mesoamerica in his notes. Rosaluna remembered that when her old uncle read his travel stories and used the term Mesoamerica, a mischievous smile would cross his face. He was thinking of the irritation it occasioned to his geography teacher in fifth year gymnasium who distinguished South, North and Central America and would not have recognised that term. Any self-respecting travel writer, however, knows that this melting pot of Aztec, Maya, Olmec, Zapotec and Kuna civilisations is a unit difficult to divide. How can Mexico be considered North American?

Another smile appeared at the word *Carib,* which Vieri used for Garifuna Creoles of British Honduras, for ex-pirates of the great gulf, and ex-Negro slaves escaped from cotton plantations of the Mississippi.

Rosaluna's curiosity was aroused when she read the notes and heard a story that Vieri told grandfather Edoardo Quirino more than half a century earlier. The notes dwelt on details that had been completely omitted in the tales, and every topic became stranded on three points: Miss Fisher, the Maya city of Quirigua, where the Fishers had a large banana plantation, and Xaman-Ek. Rosaluna was intrigued by what he wrote about the Palace of Masks, a ruin in the jungle, off the beaten track, and a series of tunnels forming an underground city, which Vieri called the City of the Howling Monkeys. She began to imagine and to go back in time.

Thirty years earlier: as usual, she was late reaching the great ancestral house at Rapolano Terme, where uncle Vieri lived. She rang the brass bell on the walnut door twice, as was her custom.

"Happy birthday, Rosaluna!" exclaimed the warm voice of uncle Vieri. He was sitting in his favourite crimson and purple velvet armchair with an adventurous travel book from his large library. The book was Conan Doyle's *Un Mondo Perduto* in the red binding of the Sonzogno World Romantic Editions.

The library was also the dream of Indaco de Lunas. Perhaps what Rosaluna remembered was his description. As a boy he had spent whole afternoons in the library. The shelves with their volumes well kept behind the hand-worked glass of the doors were in solid cherry.

The subject headings were displayed on a stand near the door: engineering, dictionaries, fiction, homeopathy, travel and adventure.

Vieri's elmwood Imca radio was tuned to Toulouse and a bright Irish rug covered his legs, trailing onto the floor. He was smoking an oval Xanthia and asked Indaco to pour them both a *jerez* from a Moser Bohemian crystal decanter in the cherry cabinet to his left. As they chinked glasses, he said "You know, Indaco, there's a mystery in this lost world of Yucatan . . . I don't understand . . ."

Caya Luna

Everyone knew that old Martinez was a good-for-nothing, a miser and a gossip. Malodorous scum had washed up in front of the old wooden hotel, if a two-storey shack with verandah on the beach of Caulker Cay with more broken tables than good ones could be called a hotel. But the lobster was fresh and Martinez's rum punch was a delightful concoction of rum and juices of papaya, mango, pineapple, carrot and grapefruit. The carrot was the secret of the drink, for which Martinez had had a poster printed: it showed a pretty mestiza with a pink hibiscus in her hair and a low-cut white dress with a fluttery skirt. It looked like an ad for Campari Cordial or a great cognac in a twenties issue of the Touring Club Italiano magazine. He had 72 copies of the poster printed: his megalomania might have exceeded this but not his purse.

Caulker Cay had eight jetties in those days, and a good number of boats. A solitary pelican was fishing between the jetties. Further out, waves boomed on the reef. Along the paths, it was common to see emerald-green hummingbirds collecting nectar from flowers. Caulker Cay was to play a big role in this story. Here lived Gamusa, a great friend of uncle Vieri.

Going N-NE along the southern Caribbean coast of Yucatan, one reaches the archaeological city of Tulum perched on a cliff top overlooking the ocean, where nature uses her most beautiful colours. The turquoise sea merges into emerald, and azure into faded blue. Federico

Fellini and Milo Manara both drew inspiration from this place. There are a thousand legends and stories in Maya culture about Tulum: stories of jungles and maize, of escapes and battles, like those described in Vieri's notes on the old capital Mayapan, destroyed in a war.

Offshore, the cays graze the surface and from the air look like rings fallen into the ocean. Gamusa knows these islands well. His boat has been round all of them, but the most fantastic stories come from Caya Luna. A grandfather of Indaco de Lunas's grandfather was born on that remote island.

A violet garden

Rosaluna day-dreamed with her eyes shut or open. These dreams came to her in the violet garden of Pacina as she lay crosswise on the Amazonian hammock or astride the eleventh branch of the old magnolia. Rosaluna could go back in time but nobody knows if she took part directly in her dreams or whether the dreams were just a product of many family tales. The indeterminacy of scientific knowledge has much to do with the passage of time.

In the garden at Pacina, the flowering wisteria coloured the ivory-ochre stucco of the back wall of the villa with mauve. The outlines of the arches of what had been the monks' walk on the first-floor verandah were visible on the wall. Before the year 1000, when the villa was still a convent, the land-tilling monks did not have the convenience of a cloister. They built the convent with a verandah that stretched the whole length of the building, which was more than 70 metres: more than enough for reading and peripatetic considerations. By about 1400, the villa was already in its present form. In the 1700s, a linden, magnolia and camphor laurel were planted, and in the 1800s the garden was closed to the north by a large *limonaia* designed by Fantastici, to the east by an informal wood where golden orioles nested and to the south by stables.

On the afternoon of 23 June 1930, Edoardo Quirino invited about 20 people. Tea was served by grandmother, aunt Rosina and Teresina under the arbour. Teresina was the maid of Edoardo's beautiful wife, Elena. For the occasion, the Hungarian gold and burnt sienna Herend porcelain tea service was used with the Sheffield teapot, and there was also vinsanto from the ten-year-old cask. Edoardo Quirino set up the radio he had built himself in the garden, with a long extension cord.

They were awaiting the return of young Vieri from his journey to Central America. For the first time, a member of the family had had the honour of being in the news: Vieri had played a major role in the business of the two canals between the Atlantic and the Pacific. Elena's cousin, Cardinal Valeri, at the time *apostolico nuncio* in France, had praised the young rascal, as Vieri was known in moments of particular affection and family solidarity.

The radio broadcast news of the landing of the dirigible, carrying the party of Italians, in Pisa. Dide, the driver and handyman, had been dispatched with the brand new V8 Dilambda to collect Vieri. The radio broadcast a second news item on the Panama Canal, without mentioning the disastrous alternative of using the Rio San Juan and Lake Nicaragua to join the two oceans. It would have been an ecological disaster for the richest tropical wetlands in the world, home to American purple gallinules, quetzals, tapirs, toucans, parrots, hummingbirds, monkeys, jaguars, pumas, wildcats and the strange masked mammal that the Indians call *pizote*, not to mention herons, pelicans, ducks and other water birds. The fact that the radio reported on Panama without mentioning Nicaragua indicated that Vieri must have succeeded in blocking the shady proposal of the competing canal. Grandfather Edoardo Quirino proposed a toast.

The radio broadcast the sporting news. Genoa was even with Sampierdarena and Andrea Doria had toppled Milan. Binda had been overtaken in the first mountain leg by a young Tuscan cyclist, Gino Bartali. Dide, grandfather's factotum (real name Alcibiade Arguti), was a great cyclist and used to say that the young man would go far. He had seen him racing on the dusty curves of the Casentino, steep and crooked like the Tuscan cyclist's nose. He added, "His nose is as sad as an uphill climb and his eyes as merry as Italians on an outing; someone should write a song about him."[2]

Tina and Gamusa

At Caulker Cay, four frigate birds were soaring high on an air current. Suddenly one dived on a young seagull and stole the fish from its beak. This is why the Caribs call them pirate birds. That morning Gamusa's boat, *Frigate,* was tied up at the jetty. Gamusa was sitting at Martinez's bar with Tina, sipping a rum punch while she drank a piña colada. Their hair was completely matted with salt, adding special charm to their not inconsiderable beauty.

Gamusa was a typical Garifuna Creole, magic selection of that mixture of Spanish conquistadores, Negro slaves escaped to the meanders of the Rio Dulce in Guatemala or to Livingstone, and Indians from the Maya mountains. He spoke a mixture of Garifuna dialect, sailor's English and Mexican Spanish. This mix had a soft but strong sound, a metallic melody interrupted by the rhythmic percussion of spontaneous laughter, winks and vivacious facial expressions. Tina was utterly infatuated. For an English girl, on that island for three years under who knows what circumstances, it must have been the love she dreamed of in some room in distant England while reading Conrad or Stevenson. Her face contrasted sharply with the Carib hair and clothes. Her wire-rimmed glasses, delicate profile and smile were those of an English girl of aristocratic family. The way she used her hands and drank the piña colada betrayed education in an exclusive college and her accent was that of an Oxford or Cambridge graduate. She wore a ring with a Welsh emblem and a finely wrought medallion that may have belonged to an Indian maharani. People knew only that she was called Tina, she was born in Cardiff, and she had come to the Americas to study medicinal plants.

Gamusa gave little away: few words and many laughs. Asked who she was he would reply, "Tina is Tina, no worries, she will always be

[2] The song was Bartali by Paolo Conte.

Tina." The islanders said they were great lovers and every so often disappeared on *Frigate* for three or four days to make love on lonely beaches. It was said that they went to Caya Luna and to that scrap of beach with six palms known as Goff Cay, and even beyond Terranova to the great blue hole.[3]

With a little nostalgia still in his cheerful mien, Gamusa said to Tina, "It's 6 p.m. in Italy. Vieri should nearly be home by now. Thanks to the two of you, the Rio San Juan is safe. In three days we can go to Rio Dulce and then to collect plants for your research in Nicaragua and Costa Rica." Tina smiled. She knew she could count on Gamusa.

The slow pace of the Caribbean was all around them. Four children were dancing on the beach to the sound of imaginary music. Another was up a palm, knocking down coconuts. Martinez's youngest daughter, a massive girl, was singing opera arias in her beautiful contralto voice. When time sets the pace of life, only fundamental things apply.

Vieri was so brown as to be almost unrecognisable. A gaggle of cousins and girlfriends crowded around him, to the irritation of Edoardo Quirino, who was impatient to have news at first hand. With his blue eyes and long fair hair, the elegant young Tuscan was usually mistaken for an Englishman. Now his dark complexion was evident, baked by the sun into a warm colour between amber and milk coffee.

Within the family, that hidden quarter of gypsy blood in the genealogical tree was acknowledged with a mixture of pride and uneasiness; the rest was pure Etruscan. Great-grandmother Filomena came from an area of the Valdichiana where hundreds of families carry the same strange surname, with zeds and a hard t, as in the word *tzigan*. Over 900 long years, that surname became Tuscan; the ancestors of Filomena included a general, a famous theatre director and that pugnacious Dominican friar, Benedetto, who took Savonarola's place at the convent of San Marco in Florence and followed his predecessor, starving to death in a cell in Castel

[3] The Great Blue Hole in Belize was explored by Jean Jacques Cousteau in 1960. It is a perfect circle in the ocean having a diameter of more than 200 m, surrounded by a coral reef that grazes the surface. It is a three-hour sail from the Turneffe archipelago also known as Terranova.

Sant'Angelo at the hand of the Medici pope. The gypsy roots were the true meaning of that strange surname. They showed in the lively black eyes of many of the girls in the family, their black hair and often olive complexion. Vieri was an exception – melanin is almost always dominant in the transmission of hereditary traits.

After about an hour of toasting, congratulations, embraces and adoring glances from the girls, Edoardo Quirino took Vieri into his study and had him tell all the details of his story, including business. He had ordered a lot of mahogany timber, a wood then abundant in Central America but rare and costly in Europe. He was happy to learn that a brig would be berthing in Leghorn in eight days time with three times the amount of timber, bought at a very good price. Besides mahogany, there was also cedar and an American novelty, chewing gum.

Vieri brought presents for everyone: jade from Guatemala, colourful naive art from Rio San Juan, and some Spanish and pre-Columbian antiques from Antigua. Later that evening he showed everyone the photographs he had taken.

Guatemala, mon amour

Sir William Fisher was fond of game fishing and archaeology, or rather, he liked filching rare articles of Maya archaeology, especially jades. The fact that visitors can now admire a necklace of cylindrical and parallelepiped beads of transparent pale green jade in the small museum in Tikal and a finely crafted dark jade medallion in the tiny museum at Chichicastenango is because Sir William's conscience began to trouble him in later years and he had his collection sent back to its sources, secretly, through certain archaeologist friends of his family. Apart from the fact that it was donated, Fisher's collection would have been the envy of any museum, and went well with the Indian jewellery the baronet had brought back from India in 1875, after serving the British Empire faithfully for more than 40 years. In India, a beautiful widowed maharani who had fortunately escaped the rite of *sati* was hopelessly in love with Sir William, and he with her, and it seems that Fisher did not return to Europe until after her death.

However, his native Cardiff was too small for the old Welshman, who then set out for the Americas, where he established thousands of hectares of banana plantation between British Honduras and Guatemala, along the road to the ruins of Quirigua. There he conceived his passion for Maya archaeology.

The immense family fortune was now in the hands of his only granddaughter, the young Miss Fisher, who lived at Castillo San Felipe, on the Rio Dulce, in an enormous wooden mansion. Miss Fisher lived a very secluded life. She was known as a genteel but severe English lady, sure to remain a spinster. Few had ever seen her and nobody had seen her hair, which she wore bound up, somewhat in the manner of a nun.

The old house of the Fishers was admired by many and feared by all. There were pirate stories and it was said that the ghost of Sir William wearing a jade mask haunted the house. It was whispered that the famous ghost ship, the pirate ship without sailors, occasionally berthed at the jetty of the house, and more than one local person swore to have seen a male figure leave the house and board the boat, which would disappear into the night.

Livingstone, a town founded by pirates and escaped slaves, can be reached in a few hours by boat along the Rio Dulce from Castillo San Felipe. For centuries, British Honduras harboured pirates in England's service. It seems that an ancestor of Sir William Fisher was one such gentleman and that he came from Cardigan on the Welsh coast. The Gaelic pin that Sir William used to fasten his shirt must have belonged to his pirate great-grandfather.

Rosaluna had acquired Uncle Vieri's love for Guatemala and his curiosity for its strange tales. It was he who gave her that pin. How it came into his hands is a mystery.

Uncle Vieri told Rosaluna about Maya cosmogony and stories about Lake Atitlan, which according to him was the only indigo-coloured lake in the world. In the evening, white clouds and mist would descend from the volcanoes onto the calm water of the lake. At dawn the shores would be brightened by amaranth butterflies and multicoloured hummingbirds.

Little Maya women with variegated violet clothes would come down from Santiago to San Antonio with baskets of fish to sell. An old man would be selling jade and obsidian. In the evening, the lake would again be suffused with indigo.

Rosaluna had dreamed enough. She wanted to know more about Gamusa and Tina, the Fishers and Yucatan. She saddled Altair and left for Chichicastenango on a freight ship. She at least wanted to see what Guatemala was like in the new millennium.

Refined connoisseurs of handcrafts say that the most interesting Guatemalan market is that of San Francisco el Alto; however, the most beautiful, intriguing and stimulating is certainly the Sunday market in Chichicastenango. Arriving on a Saturday afternoon, one can stay in a hotel which was formerly the home of the Spanish governor. The rooms, furnished with Spanish antiques, open onto two patios and guests have a native manservant who discretely enters the room at night to tend the open fire.

In the evening, many fires are lit in the market square because most of the vendors have already arrived from the mountains and nearby villages. There are peasants, artisans, shamans and medicine women. Syncretic rites begin on the steps of the two churches. Clouds of incense blow from strange contraptions and young girls dressed in white ask the Virgin Mary and the shaman for a happy marriage and children. Roasted cobs of corn and small salted fish are eaten. The atmosphere is dense with smells and sensations; mystery and presence give this poor population dignity and richness.

On Sunday morning the square swarms with people and stalls. There are brightly coloured fabrics, candles of natural beeswax, musical instruments, wooden masks, silver filigree. The vendors are never insistent. On the nearby Sierra Lacandona, at the Mexican border, young Indians, always in conflict with the various governments, plot revolution and release from agelong poverty.[4] The market in Chichicastenango

[4] In recent years a new revolutionary movement has arisen in Chiapas, the region of Mexico bordering with Guatemala. The movement took the name of Emiliano Zapata who was from the San Cristobal de las Casas area.

with its healers and incense is wonderful and magical, the heart of the Guatemalan highland. The hummingbirds are minute like the fuchsias and orchids from which they steal nectar.

Rosaluna went into the church in the market square. An old Indian woman saw the Gaelic pin and spoke to her. The pin had probably belonged to a "gentleman of fortune." The old woman confirmed that the ghost ship that sailed up the Rio Dulce on spring nights belonged to another "gentleman of fortune" by the name of Gamusa. She also said that Gamusa's mistress was Miss Fisher, alias Tina, and that they and another Italian by the name of Vieri, to whom Sir William Fisher had given the Gaelic pin, were the only people to know the secret of Xaman-Ek, the pole star, and to speak the language of migratory birds and to know the way to the fourth unknown Maya site, guided by another star that the Arabs call Altair. In this site there is no treasure but ancient manuscripts of the Popol Vuh that describe how to live in harmony with nature.

There was a strong smell of beeswax in the church. A family of Maya from the mountains brought cobs of maize to the priest, interrupting the mass. The priest embraced the donors and the syncretic rite resumed, with renewed impetus.

Rosaluna mulled over the last words of the old woman, "To combine Xaman-Ek and Altair, the calculations of a mathematician from St. Louis, a stormglass and the bird migration routes are needed." Also, "At first we isolate beauty from its context but in the end we understand its harmony with everything else."[5]

On her way home she stopped at Antigua to see the ruins of the ancient capital, destroyed by many earthquakes. One of the churches reminded her of San Galgano, near Siena, with a roof of sky and stars. In the afternoon she was caught in a tropical downpour which soaked her to the skin. She sought shelter in a café where she had a Zacapa Centenario to warm her up. She opened the book by Rigoberta Menchu

---

[5] Rabindranath Tagore.

that her grandfather had given her before her departure. Her imagination was full of Rigoberta's stories and struggles and the story of how she was awarded the Nobel Peace Prize. She remembered that her grandfather's dear friend, Antonio Melis, full professor of Hispano-American literature at Siena University, had played a major role in making Rigoberta known throughout the world. His wife, Lucia, came from Rapolano Terme, like Rosaluna's great-grandparents. When she looked up, she saw Antonio and Lucia seated two tables away.

The strange thing was not only the probability of being in the same café in Antigua on the same day but that when Rosaluna looked up she knew that Antonio Melis was in the room. Now she began to understand. The pieces of the puzzle were falling into place: relations, signs described by Novalis, the importance of roots and family traditions, the Gaelic pin, Altair, the sacred books in the Mauritanian desert, the stormglass and the pelicans' routes. However, most of all, as her grandfather had taught her, the paths of knowledge are never solely rational or solely spiritual: a sort of syncretism of emotions and affections on one hand and scientific logic on the other.

She sat on a bench in the dilapidated church and felt her love for Sinbad, for all beauty, for dreams that come true, welling ecstatically in her breast.

6

Belisario's rum punch and more about the bird islands

To Tito Lucrezio Caro

The rare wooden tobacco duck
was engraved by a fine sculptor,
you can see the craftsman loved it
and perhaps also its destiny knew.

Around the big ancient fountain move
the ducklings of the mandarin
they've an old wide-duck as a friend
that got lost migrating to the sea.

In the wings colours you can see
an arabesque design of Greek frets;
in the flowers the same identical frets
telling a story of forms.

In Inclination Kandinskij draws
the story you've already sung
talking of Clinamen, evolution,
and of our destiny already told
in the molecules of our life
with coloured dances in the Chance's hand
in the dialectic infinite story
that ties choices and fate in a vase.

The vase that contains the biosphere
and the structure connecting the world.
Maybe, Lucrezio, that utopia of yours
is, of the same essence, the round-dance.

Rosaluna returned from her first trip to Guatemala with mixed feelings. She looked forward to returning home and talking to Sinbad on the telephone but she felt impatient to start on the path indicated by the Maya woman in the church at Chichicastenango and she was curious to explore how her life and that of her family were linked to those stories, apparently so remote in time and space. In the end, wisdom prevailed. She was not yet ready for a second trip to Yucatan. Indeed, three years were to pass before Rosaluna, this time in the company of Sinbad, returned to Guatemala.

During lazy hours in the garden at Pacina, broken by gallops with Altair and a few visits by Sinbad, Rosaluna decided to spend some time at Rapolano with uncle Vieri, before he lost his proverbial lucidity. She wanted to learn how to travel, how to manage difficult situations, how to read signs that we often do not recognise. Another thing she did was to ask grandfather Veliero to continue his story about the great migration of the birds.

A contralto voice

The art déco poster with the pretty girl in bathing costume announced that Belisario Martinez rum punch was "famous in the Caribbean and elsewhere." The punch was made with seven juices as well as rum. Rum and ice, Martinez would say, "If you don't have Zacapa centenario or Matusalem or Barbancourt, Bacardi will do, but there must be seven juices: mango, pineapple, pear, grenadine, sour cherry, orange, grapefruit and don't forget the carrot."

Belisario's daughter, Francine, was granted two, or rather three, things by Mother Nature: a splendid contralto voice and a prosperous bust. She attracted men like a mermaid, singing Carmen and then enveloping them in her hazelnut-brown feminine abundance. Passing sailors declared their satisfaction and often called in on the next voyage. As a result, Francine had three little boys. Yanez was the son of a Portuguese, Aramis of a French sailor and Duvalle issued from Francine's unlucky love affair with a Garifuna Creole from Dangrija, whose boat was lost in a tropical storm on its way to Barbados.

Belisario, together with Francesco and Gamusa, had contributed to the success of the birds' new islands. His experience with the hurricane house had led to special know-how in the use of timber and mortar.

Pearls and manatees

At the archipelago of the 11 islands, the construction of houses, storage, nests and shelters was now complete. Incessant work by the albatrosses, pelicans, manatees and other birds had brought the dream to fruition. Remote consultancy with Belisario and several sea voyages with Francesco and Gamusa had provided other materials to complete the birds' village. At the big celebration, at which Francesco and Gamusa represented the human nation, a "trade agreement" was signed and is honoured to this day in those happy islands. Francesco and Gamusa agreed to bring things required by the bird nation twice a year: new seed, mud for construction, timber, and new flocks of birds wanting to joint the distant colony. Once they brought seven pairs of Gould diamond finches, a pair of scarlet rosefinches, lanceolated warblers, a flock of laughing doves, a family of waxwings, golden orioles. In the meantime, the beautiful blue rock thrush and blue-cheeked bee-eaters had had extraordinary reproductive success on the islands.

The trade agreement was based on the fact that the food chain of the island where the pelicans and manatees lived produced a large quantity of pearls. These were given to Gamusa, who used them to build a fine brig for his trips to the 11 islands and other "incursions." The manatees saw their islands come to life, with the great input of bird fertiliser and bird society, and helped the pelicans find oysters. With the help of other fish-feeders, who greatly appreciated fresh oysters, they brought the pearls to shore in their great beaks.

In the evenings, Polybio read *Gargantua and Pantagruel* and the adventures of Talamega in the Sea of Chaos to the birds.

The Madeira petrel

Only 30 pairs of Madeira petrel (*Pterodroma madeira*) exist today. They live on Madeira and are regarded as the European bird closest to extinction. One of the battles won by Gamusa and Francesco was to prevent erection of a Portuguese military radar base on Madeira. The two friends sent so many birds from the 11 islands to disturb construction work that the government decided that the site was unsuitable because of migrating birds. Once the aim had been achieved, the migrations disappeared as if by magic and the petrels continued nesting.

Once the boat of Francesco and Gamusa assisted the production of the wonderful film by Jacques Perrin, *Winged Migration*.

Wallace's psychedelic dream and biological evolution

One evening, after Polybio had finished reading the masterpiece of Rabelais (*Gargantua and Pantagruel*) to the bird nation, he explained the importance of biological evolution and the role of beauty in the history of nature. Polybio gave a full university lecture.

"*Sacaleli* is a word used in one of the many New Guinea dialects to indicate a nuptial dance performed by 20–30 males in brightly coloured livery. It is danced for a group of drab-looking females. At the end of the dance, each female chooses a temporary companion. The dancers are the males of *Paradisaea minor* or the minor bird of paradise. They become so absorbed in the dance that nothing can deter them or scare them away.

Alfred Russell Wallace must have seen the dance and was the first scholar to gather data on this wonderful bird. Wallace was a very creative man. In 1856, several years earlier, he had a psychedelic experience while delirious with malaria in the jungles of Ternate, Indonesia. This experience revealed the essence of biological evolution to him and prompted a famous letter to Darwin which became a milestone of modern biology.

Wallace would not have made the mistake that Bateson calls the *antiaesthetic assumption*, derived from the importance that Bacon, Locke and Newton attributed to the physical sciences, namely that all phenomena can and must be studied in quantitative terms.

Wallace divided the islands of the Sunda using aesthetic criteria. The Wallace line divides Borneo, Sumatra and Java from Celebes, Timor and the Moluccas on the basis of the different aesthetic characters of races (Malays and Papuans) and the different flora and fauna. To Wallace, anthropologist, traveller and biologist, we owe dictionaries of 75 dialects of this Asian archipelago.

The role of aesthetics in biological selection was clear to Wallace, who while exploring luxuriant jungles on a remote island, discovered birds of paradise. He pondered the ages and generations that it must have taken these creatures to evolve, in a country where birds of paradise hatch, grow, live and die, year after year, in the depths of the jungle, far from any "intelligent" eyes to observe their splendour. He marvelled at such a waste of beauty!

Birds of paradise, with their multiform colouring, their variety, are a good symbol of aesthetics in nature. *Paradisaea apoda*, discovered by Wallace in the Aru islands, is a living pattern of pastels and colour combinations of rare beauty: deep yellow, burnt Sienna, purple, chocolate brown, golden green and grey-blue. *Paradisaea sanguinea* or the red bird of paradise shows off its colours as if it were aware of their effect: lacquer red, grass green, orange-yellow, brown. *Xanthomelus aureus*, or golden bird of paradise, was already rare when Wallace discovered it in New Guinea – this bird with golden plumage was believed to feed on dew, nectar and light, sprinkled with nutmeg."

Lights in indigo

Not long after this, a few days after the tsunami, Francesco, Gamusa and Polybio discussed what had happened in the Gulf of Bengal. It was New Year's Day 2005. In the archipelago of 11 islands all was well, but a tragedy had occurred in the Indian Ocean. The friends made two important observations: no deaths had been recorded in the national parks, despite the large population of wild animals, who had evidently felt premonition or detected the seismic waves and had sought refuge in the woods and mountains. On the other hand, domestic animals had behaved differently and had died in their thousands. The three agreed that a "low tide" exposing hundreds of metres of sea bottom is an exceptional event, a phenomenon showing the great force of nature. In such cases, wonder should be accompanied by fright, an edifying emotion, as well as flight from the sea shore. Nature is neither benign nor cruel, but may show her power. Humans living in an increasingly virtual world forget that they belong to this beautiful but terrifying planet.

Francesco took a stormglass out of his pocket and presented it to Gamusa "to see storms before they arrive." Then he warmed a shiny black stone in his hand, a volcanic stone from Kashmir, a gift from his dear friend Abdul of Srinagar. Francesco shook the warm stone and held it to his ear to hear the sound of the water imprisoned in it.

"Humans should listen more to nature," he said.

Gamusa reminded Francesco of when the Indian Ocean had taken the body of Sheshadri, a physicist who studied time and evolutionary equations of the structure of matter, at Madras University, friend of Indaco de Lunas.

Professor Vatsala, director of the Centre for Biology of Complex Systems of the same university had invited Indaco de Lunas to give a seminar. The letter was delivered to his department at Siena University. Professor Vatsala regretted that Sheshadri could not sign the invitation personally because he had disappeared three months earlier. He had disappeared in the waves of the Indian Ocean, in front of his house on the coast near Mahabalipuram, south of Madras. He did not leave clothes or other traces. The ocean did not give up his body. His scientific imagination was probably lost in the whorls of shells and the elegant structures of seahorses. Perhaps the ocean had taken the intuition of a great scientific discovery: the possibility of conjugating the awesome aesthetic complexity of nature with thermodynamic functions, entropy and time in evolutionary physics.

The Centre for Biology of Complex Systems is surrounded by fragrant sandalwoods and brightly coloured flowers. A lady wearing a sari of the same vibrant colours greeted the Italian visitors at the door and invited them to remove their shoes. The Italians followed her into the director's office, curious to meet the writer of the letter. Smiling, she took her place behind the desk and welcomed them: she was Professor Vatsala.

Sheshadri's widow was at the seminar that afternoon. They embraced: she had come to Siena a few year earlier with her husband. She felt that Sheshadri was living on, somewhere. After the seminar, Indaco lingered with Mrs. Sheshadri. She did not mention her husband's disappearance; it was not real to her. They spoke of his desire to do real physics, *physis*, the passionate science of nature as she really is. Mrs. Sheshadri recalled her time in Siena, the mysterious words in the magic square on the external left wall of the cathedral.

S A T O R
A R E P O
T E N E T
O P E R A
R O T A S

At the Madras research centre they had a colour, indigo, that they produce by ancient processes. Marquis, a friend and colleague of Indaco, is examining the plants, looking at seeds and has them give her a small chalky fragment of a colour between azure and violet, the mythical indigo. It is the colour of the ocean that took Sheshadri. On the other side of India, in Kerala, the indigo of the sea at dusk fills with thousands of lights. Those of the boats form a great floating city with the consistency of mercury on the indigo background.

On their return to Siena, Indaco and Marquis plant the seeds near the fountain sculpted by Pietro Cascella, next to the butterfly shapes of the water jets, bringing art and science together in a garden shaded by the third circle of the town walls!

Conversation with the two Indian ladies shifted from Siena to Jaipur and the artistic and scientific stature of the Arabian architects of the city in Rajasthan, the pink tones of the astronomical observatory, the enchanting Wind Palace and the Venetian beauty of Jaisalmer, a desert town on the caravan route.

Two days later, Indaco and Marquis may have been speaking too loud over their Darjeeling in the cafe of the Maharaja's Palace in Udaipur. A Japanese gentleman at a nearby table overhears them and with impeccable courtesy engages them in conversation. He knows Siena and wants to speak about colours and forms. A coincidence? Stochastic encounters? He is an artist who paints and draws only with indigo and who transforms natural earths and plant dyes into art.

A few hours later, in an Udaipur shop, Indaco de Lunas sees miniatures on metal boxes that recall Sienese encaustics. Thus, Indian alchemies encounter stories of Siena and tell of a "time scientist" lost in the Indian Ocean, south of Madras.

7
Tales of Uncle Vieri

Tikal

From the nectar of the fuchsia
the hummingbird sucks energy
and by ancient alchemy
transforms the sugar
into vibrations of force.
Energy in the far infrared
excited by sunlight
that the flower hid
and now the whirr of wings.
The old Guatemalan observes with child-like wonder
a story written long ago in the verses of the Popol Vuh.

The previous year, Rosaluna had climbed the travertine steps of the old tufo house at Rapolano Terme. She brought a gift of 101-year-old balsamic vinegar from Pacina in a crystal vial. The Tuscan gentleman, now over 80, was sitting in his threadbare crimson and violet velvet armchair smoking an oval Turmac. He offered Rosaluna a drop of ratafia and granted her request to set aside three days to tell her the stories of his adventures as a young man in the 1930s.

"Some of the stories are by Edoardo Quirino, others were written by me with Vieri protagonist in the first person," he explained. "Not all the stories are about the characters that interest you or about Yucatan."

Rosaluna replied that she was happy to listen to stories that had so fascinated her as a girl. She considered that knowing about different experiences from other periods, difficult travelling situations, would be a school of life. She recalled that a Caboclo in the Amazon had quenched Vieri's thirst with water obtained from a tree and a Seminole had started a fire using two knives. Vieri smiled and began with Incense Smoke at *Chichicastenango* in Yucatan, where Rosaluna had been.

Incense smoke at Chichicastenango

The steps of the church were saturated with wax, incense and plant essences. The whitewashed façade contrasted with the greasy stone of the steps. A Maya shaman was dispensing blessings and incense with archaic priestly movements, in front of the closed door. Night had fallen several hours ago and in the square of Chichicastenango, many people slept or camped around charcoal fires, on which they had cooked tortillas and offal. In the morning it would be market day and only the rare insomniac still walked the streets.

As Vieri slowly climbed the steps of the church, a young couple appeared from the other direction. The young woman had a very small baby slung from her shoulders in a red wool shawl. Even in August, nights are cold in Chichicastenango. The two muttered to the shaman in low tones and began lighting candles in the parvis. The incense spread densely in the damp night air. The mother hobbled three times around the parvis on her knees, until they hurt. Then the couple kneeled together and the shaman spoke in a strange tongue that was neither Spanish nor Latin. Several times he placed a paper cone, which may have contained ancient prayers or relics, on their heads. The light of the candles lit the shaman's features from time to time. He may have been fifty or a hundred: beautiful dark eyes, high angular cheekbones, a dusky weathered face, the face of an old fisherman transposed to a mountain village. A scar under the left eye told Vieri it was Miguel Guarcas, the man he had come to meet.

Vieri waited until the couple had gone; then, with a hexagonal 50 quetzales-cent coin in his palm, as instructed, he asked in good Spanish, "Have you any fish for sale?"

"Lake fish and river shrimp," was the reply.

Several days later, reflecting on the magic days spent at Chichicastenango, Vieri had trouble focusing on his sensations. Later he wrote in his travel journal about being "immersed in a medieval atmosphere," adding that if a people has time and money for flowers, then their civilisation is alive and present. Rereading the diary 70 years later, one finds pages of great scientific interest. Vieri writes about a hummingbird on the patio of the old colonial hotel in Chichicastenango. The bird dipped several times into the nectaries of some fuchsias. Vieri digresses in a comparison of hummingbirds and sloths. His observations concern the role of entropy in nature, that physical property of matter by which everything except life degenerates into disorder. Vieri observes that sloths and hummingbirds have solved the problem brilliantly in two quite different ways. The sloth survives by being a slow and careful climber of tall Amazonian trees, where he evades the ocelot, his natural predator; the hummingbird by lightning visits to flowers. Both transform solar energy trapped in the highest shoots of trees and in fuchsia nectar; both oppose entropic degradation by strategies based on solar energy.

Miguel had given Vieri an appointment next morning at mass in the church of St. Tomas. In a country where syncretism reigns supreme, the catholic priest or shaman transmitted precious information. Vieri was also given the notes of a 16th century Sienese monk from the convent of San Bernardino of Siena in Valladolid. The information was about Don Juan de Grijalva, the first European to see Tulum, and was to prove useful in subsequent journeys into Yucatan.

Many years later, three friends met in Vieri's study at Rapolano Terme to talk about Don Juan de Grijalva. Vieri ushered them into the dark study and then lit the four brass chandeliers made with old acetylene tubes according to designs by Edoardo Quirino. The styles of *art nouveau* and Nicaraguan naive art from when Vieri was in San Juan were elegantly combined in the four chandeliers, without losing the softness of the pomegranate flowers and hummingbird wings, which were the main motifs. The chandeliers were similar but different, with the arrow to the north, the spiral to the south, the moon to the west and the sun to the south. The symbols were obvious but not banal.

Vieri took two bottles for great afternoon occasions out of the cherry cabinet: sweet liqueurs suitable for gentlemen of a certain age – anise and ratafia. Vieri, who considered alcoholic infusions of anise and cherry to be one of the pillars of Mediterranean civilisation, actually showed them the whole collection of anises, from Spanish Machaco to Pernod, Ouzo and Meletti anisette from Ascoli Piceno. They all soon had a frosted glass with anise and ice. Some smoked a Reconcavo, others a Gitane. Vieri puffed on his oval Xanthia. They relaxed in the comfortable crimson and violet velvet armchairs and Vieri asked about the scientific memoire on the Madeira petrel presented by young Indaco, as junior member of the International Society of Zoology, at the world ornithology congress at the abbey of Pont-a-Mousson, near Nancy.

"A great success," exclaimed Vieri enthusiastically.

"Claro que si'," added Gamusa, as Vieri continued to nod.

"If I'm not mistaken, it is the first time that a 15-year-old has ever spoken at an international scientific meeting," said Vieri.

"Not exactly," said Indaco, "because I turned 16 on 4 February and conferences on mathematics and logic accepted memoires by boys of 14 and 15 last century."

Gamusa was proud of his young friend. Now he had a *papito* and a younger brother. He had continued to call Vieri *papi*, the big father with grey hair, since the latter's first visit to Caulker Cay five years earlier, and now his *entente* with Indaco de Lunas was total: in tastes, adventures, daydreams and ideals. Indaco, Vieri's young nephew, son of the gypsy from Granada and the cousin of his Sienese friend, appeared suddenly in Gamusa's life like a frigate bird representing the certainty of friendship and knowledge of the past. The geography, anthropology and natural sciences that Gamusa felt and knew, as an Argentine sailor knows the Southern Cross, as an Amazonian Caboclo knows how to drink from the water tree and as a little Maya knows how to heal with herbs, the sciences that Gamusa loved without having read, had come to fruition in a complex way in young Indaco, through study and experience, intuition and enthusiastic desire, which were a bastion against the arrogant approximation of the sorcerer and the smug notions of the specialist.

1930: The Colonial Stores, Antigua

They disembarked at the *Colonial Stores* in Antigua. They had reserved almost the whole floor of the old hotel: rooms, a big, well equipped room for communications, another as storeroom, and the suite, where they set up their base for the various expeditions. Experts in Latin-American zoology and botany, jade experts and oceanographers came from the University of Mexico City, from Costa Rica and British Honduras and from the publisher Fondo de Cultura Economica, Mexico City. The hotel belonged to an Italian, a genial Romagnolo by the name of Guido Pasi da Alfonsine. Pasi had transformed his old import–export firm, the "Colonial Stores," into a comfortable place for Europeans to stay in Guatemala. Europeans came for trade or archaeology, or in the present case for scientific and mineral research.

The group of guests was picturesque and heterogeneous. Besides our two Italian friends, Vieri and Edoardo Quirino, there was an Italian with some Creole blood by the name of Amarilli: she was doing a research doctorate in natural sciences and somehow combined striking beauty with a rare understanding of her discipline, intuition and creativity. Gamus and Tina had joined them from Caulker Cay.

The research project was called QU.AR.TO, but had nothing to do with Garibaldi's Thousand redshirts, even if the ship that brought the party to Latin America sailed from Genoa. The acronym, coined by Vieri, meant Quetzal-Archaeology-Tortuga with reference to a hypothetical "fourth" Maya site, the existence of which the Tuscan gentleman was certain. On the strength of this he had convinced a major Siena bank, an industrialist from Nuremburg and various universities to finance the enterprise.

Five expeditions were planned. Two had Costa Rica as destination: the first to the Monteverde mountains on the trails of quetzal migrations, in those days still quite mysterious. This bird with emerald green plumage is sacred for the Indians of Mesoamerica. The second expedition was to Pacific beaches on the migration routes of giant turtles. The other three expeditions were to Mexican and Guatemalan Yucatan and Gamusa's islands in British Honduras. Vieri had given orders about the Mayan archaeological sites, especially Tikal, the jade mines, the sacred *cenote* (deep tanks of freshwater, very interesting ecosystems), certain freshwater mammals, the writings of friar Bartolomeo de Las Casas, the syncretic rites of Chichicastenango and the coral reefs off British Honduras. He had chosen front-line figures in the various fields to make up the expedition, all with lively curiosity and spirit of adventure. As usual, Vieri had not revealed the final aim of the project. Curiosity was therefore at a high pitch and everyone was eager to begin and discover the end of the story. They all trusted Vieri implicitly, most of them being old friends of the Tuscan gentleman and familiar with his rectitude, culture and competence.

On the patio of the hotel, they were drinking piña colada that the Romagnolo had prepared for their arrival: the pineapple juice was exquisite, the ron (or rum as the English call it) was the very best from a clandestine distillery in Martinique, and the Guadalajara goatsmilk cream added to the cocktail was an unusual choice of the host.

Nineteen hundred and thirty was a strange year. The friends were discussing sport: in the Giro d'Italia Luigi Marchisio had defeated Alfredo Binda who had won for three consecutive years, and Uruguay had won the first World Cup at Montevideo.

"I bet Italy will win the Cup in 1934," said Vieri.

"I think it will be Spain," said Guido Pasi, "they have a goal keeper who can stop any ball. Zamora, you'll hear about him!"

The talk and bets dispersed in the merry-making and anticipation of adventure. The night before the expedition they all slept peacefully.

The enterprise turned out to be a scientific and archaeological success: interesting specimens were found and Maya symbols linked to Xaman-Ek, the Maya pole star, were deciphered.

Mullet roe and the beach of giant turtles

Vieri was known for his laziness. Most of his journeys and adventures occurred in his velvet armchair with a good book and a glass of port. This time, however, the journey to Rio San Juan and Guatemala aroused his curiosity to such a degree that he forgot his laziness and planned a second voyage to Mesoamerica within three weeks of his return.

That evening at dinner, aided by a good Trebbiano from Pacina and fine slices of fresh mullet roe with Pacina olive oil, he had made his decision. The mullet roe had been delivered in a packet from Sardinia that morning, with compliments for the Rio San Juan business, together with some bottles of white Mirto. The senders were colonel Sergio Caroli and Renzo Stefanelli, with whom Vieri had shared other adventures.

After dinner he settled in his armchair with a rare little book that was particularly dear to him, Jules Verne's *Paris au XX siècle* published in 1863 and never reprinted. Verne forecast that the demon of property would drive humanity relentlessly forward, and indeed in the epigraph he transcribed the words of Paul-Louis Courier: "Pernicious genius, devoted to the wordly sciences and vile mechanical professions! What would you not do, left unattended in your fatal impetus of inventing and developing?" The Imca radio in Vieri's study was broadcasting news of the death of Sir Arthur Conan Doyle in Sussex. It was 7 July 1930.

Vieri was an insomniac and that evening he could not get to sleep. Twice he got up, put on his silk and cashmere dressing gown and returned to the great globe in his study. He took measurements, consulted maps of Yucatan on the table, poured a glass of port, made himself comfortable in the velvet armchair and rose again to look at the globe or the maps.

"Maps and coincidences," he murmured, "Too many coincidences for my hypothesis to be imagination."

He did and redid the calculations. Palenque, the Maya city in Chiapas, was about 250 km west of Tikal. Quirigua, the sacred Maya site at the Guatemala–Honduras border, was about 250 km south of Tikal. Actually, the two archaeological sites were slightly off west and south, as were the southern and western steps of the *Mundo Perdido* pyramid of Tikal. Vieri reasoned that Mayan architects were too skilled in astronomy to make a mistake. At Chichen-Itza they built a pyramid, the corner of which reflected the sinuous figure of a serpent onto the steps at the spring and autumn equinoxes. The head of the serpent was exactly where that fascinating shadow began, only two days in every year.

That night another coincidence involved Vieri in complicated calculations in nautical miles on maps, obtained from the British Consulate by cousin Imelda who worked in the Italian diplomatic service. It concerned the "blue hole" off the archipelago of Turneffe in the Caribbean. This extraordinary circular coral formation lies exactly east of Tikal, at a distance of 250 km.

Vieri's observations in Mesoamerica kept him busy for several days and various customers of his armoury and hardware shop on the ground floor of the old tufo house complained that it was impossible to speak to him. The hunting season was approaching and Vieri did a good trade in hand-made cartridges with the powder recipe and doses of great-grandfather Guido. However, there was always someone who wanted Vieri's advice: these are good for partridges, these for wood pigeons on damp days, and so forth. There was actually only one type of cartridge and the ritual advice was always the same, but since "the owner's eye gilds the horse," Vieri had to leave his study several times and go down into the shop. The trade he had inherited was far from his liking and cultural leanings.

He poured himself a second glass of port and wrote, "All the *cenote* I visited are circular and the material surrounding them originated from coral. The *cenote* are Mayan sacred sites. The 'blue hole' is therefore a Mayan sacred site, the third indicated by the *Mundo Perdido* pyramid." The syllogism was not the best, but it enabled Vieri to smoke the last Turmac and finally go to bed.

Costa Rica two months later

Just before midnight, the fisherman called to fetch the Italians at Jim Gordon's. "The moon's hump lies to the west," said Vieri, "so it is waxing and nearly full. A good night for turtles."

Amarilli was visibly moved. She had heard that when they are 20 years old, giant Pacific turtles return to where they were born to lay their eggs. In her wildest European naturalistic dreams inspired by Darwin and Gauguin, she had never dreamed she would witness the sight. She now felt certain she would see a giant turtle that night.

They set out southward along the beach. Juan was walking fast, followed by Vieri and Gordon. Amarilli and Gamusa brought up the rear. Loud noises from the nearby jungle reached the beach: monkeys, night birds and loud cries of large predators. Juan Valdez, the fisherman, drew their positions with a stick in the sand: one person every hundred metres, or thereabouts, high on the beach away from the water. Amarilli was given the northernmost position.

An hour passed, maybe more. Amarilli had adopted Vieri's habit of not wearing a watch so as not to be a slave to time. She gazed upon the gentle waves of the moonlit sea.

Suddenly there was a great shadow, high in the sea, enormous. The shadow was approaching slowly. The giant Pacific turtle with its leathery shell was returning to the place of her birth after 20 years, to the beach where in a desperate flight for the water, she had escaped the fate of many of her siblings who were devoured by birds. She came on a moonlit night. The great shadow, nearly two metres in size, slowly approached the shore, a mountain in the water. She resembled an upturned boat, a living fossil among the waves. Amarilli felt apprehension and elation. The turtle had reached the high tideline. She made various tentative holes in the sand and in one of them finally laid 30 or 40 eggs the size of ping-pong balls.

Juan had been a deep-sea fisherman in his youth. Now, at 50, he lived by raising fighting cocks and marine turtles, the "Carey," the most beautiful and rarest of sea turtles. Active all his life in ecological campaigns for the reproduction and conservation of the green *tortuga*, he nevertheless did not disdain betting on cockfights and personally chose a silver spur from his collection. A *delicado* always hung from his lips and Teresita was always on hand.

Teresita was a green parrot with a blue forehead, named in honour of a beautiful woman of San Cristobal de las Casas, known to Juan in his youth at Puerto Escondido. Teresita was rendered immortal in a painting of the most famous naive painter of Costa Rica, a certain Xavier who lived at Puerto Viejo de Talamanca.

That night, after seeing the turtle, the party had colourful dreams. Vieri dreamed of Hieronymus Bosch's *Garden of Delights*, Juan of Gauguin's *The Poor Fisherman* and Amarilli of Diego Rivera's *murales*.

Monteverde, Costa Rica

Vieri was attracted by small forms of nature and their colours. His attention was arrested by the grey transparence of an *Epitonium* shell, the green of a tiny *Smaragdia viridis*, the movements of a seahorse or of a hummingbird poised above the petals of a flower. Gamusa had led them into the forest of Monteverde. The mists and low clouds had cleared and a ray of sun lit bushes of colourful flowers in a clearing between a brook and some rocks. In that bright space, hummingbirds of various species hung in the air like holy ghosts before dipping into flowers. They counted six or seven species of different sizes and colours. Vieri recognised the violet *Campylopterus* and the bright green *Thalassinus*. There were also beautiful *Doryfere* and *Eugenes fulgens*. Vieri let out a cry of admiration at a brilliant *Panterpe insignis* only 30 centimetres from his face. Its throat was coppery red, its head turquoise and a bright yellowish orange collar merged into the intense green of its mantle. An indigo-throated *Amazilia* sipped syrup from Gamusa's hand. Vieri's photograph of this bird later won a prize in Italy. The shutter of his Leica clicked merrily as he sought out forms, colours, movements and combinations of tones.

Monteverde was the habitat of the *sapo dorado*, the golden toad of the misty forest. However, Vieri and friends found many of these toads inexplicably dead. Amarilli explained the mystery: the hole in the ozone layer let in ultraviolet rays that damaged the fragile golden skin of the light-sensitive toad. An article in *Nature* by a scientist friend of Vieri's announced the extinction of the *sapo dorado*.

Later, on the road down towards the Pacific Ocean, the party stopped to eat in a tavern. Nearby, a potter was removing articles with a shiny black glaze from a kiln and pairs of *Ara macao* were returning to the mountain to roost. The atmosphere was right for confidences and stories, and thus Vieri came to know of the love story of Gamusa with the lady of the *lagoon of dreams*, a lagoon above Gamusa's native Garifuna village in Honduras. The lagoon is three days' walk from the sea towards the high hills of Peten, or along the coast to Livingstone and from there up the Rio Dulce to the castle of San Felipe in the jungle of Peten.

Vieri returned from Mesoamerica with useful information on the *cenote* and on the Rio Dulce and Peten area. He also deciphered a Mayan inscription about navigable subterranean canals between *cenote* and sacred sites.

Dark eyes and submarine paint

Vieri read the next story, explaining that he wrote it for a literary competition. He had imagined himself in advanced age in 1938; his nephew Indaco de Lunas was the narrator. The story had nothing to do with Yucatan.

Rapolano Terme, 1938

When I parked the Lancia Aprilia in the old town square, I was aware of the interest and envy aroused by the luxurious bottle-green enamel of the new berline. With bashful pride I quickly crossed the square to the armoury and hardware shop under the old tufo house. After a glance to ensure that Vieri was not in the shop, I ran up the two flights of stairs.

Vieri was a most unlikely shopkeeper! His mind was always elsewhere. This, however, was the trade his family had assigned him. Now, more than 50 years later and with the somewhat rhetorical wisdom of my 90 years, celebrated on 6 February last, I realise that uncle Vieri was probably the victim of the period in which he lived, or rather, of times and places determined partly by fate and partly by choice, like all the stochastic processes of life. If the tufo house at Rapolano Terme had been the famous sandstone house on 35th Avenue, New York, uncle Vieri could have been the Nero Wolfe of his time, but because crimes and misdeeds are less frequent at Rapolano Terme than in the American metropolis, uncle Vieri only made his refined contribution in a few special cases, as I am about to show. I think that if he had been born 50 years later, his stories would have been published, instead of circulating among members of the family, where they became part of a sort of *family literary history*. In this history, appreciation of ecclesiastic personalities in the family, especially on the maternal side, had a certain weight. A cardinal, Valerio Valeri, from Santa Fiora on Mt. Amiata, was *nuncio apostolico* in Paris during the controversial Pétain regime, and only just escaped being sent to the papal throne. At some point in the course of long family soirées, an aunt or cousin would usually whisper something about Vieri's friendship with a writer he met by chance in Trieste, the young Morselli, "so dark," "so elegiac," and about his escapades in that city, a cultural cauldron at the start of the century. "Who knows," they would add, "perhaps he had tea with Svevo and Joyce in some Triestine café."

Breathless, I reached the big walnut door and rang the brass bell of the Vieri mansion twice, as was the custom. Uncle was in his favourite, particularly fine armchair with a book from the Travel and Adventure section of his large library. The elmwood Imca radio was playing a fox trot. An Irish rug lay over his legs and trailed onto the floor.

"Well, Uncle?" I asked, full of curiosity. "Did you receive that letter from the Swiss embassy?"

"I received a big box of delicious candied fruit from cousin Imelda. Here, have one! Try the little green citron."

"Please, Uncle," I insisted, "I am dying to know!"

"Have one." He opened the box of candied fruit and I immediately noticed the closely written sheet of tissue paper inside the lid. "Imelda is invaluable, as usual. It was indeed the special submarine paint produced since 1899 by the firm of the father-in-law of Italo Svevo. That year," he continued, "I was only passing through Trieste."

The Lloyd Triestino cruise to the Bosphorus sailed from Trieste and Vieri had availed himself of two extra days of leave to see friends and enjoy the spring atmosphere of the seaboard city. The book he was reading when I burst in slipped from his hands. It was the third volume of Verne's *The Children of Captain Grant*; an old photograph, more brown than white, slid from between the pages onto the floor. I just had time to see the dedication written in Cyrillic and a date: Odessa, summer 1899. The pretty young face of a girl with Slavonic features, or perhaps oriental, quickly disappeared under the Irish rug into a big pocket of Uncle's silk gown.

"The *Lidia* . . . I remember the name of our cruise ship to the Black Sea because four years later my friend Morselli embarked mysteriously for Uruguay on the same ship. In Trieste, in that period, there was talk of new chemicophysical studies on certain special submarine paints: resistance to saltwater, ferroelectric and ferromagnetic properties and above all, they seemed to be invisible to ultrasound. I can understand why the German Secret Service would be interested in such research. Those paints could have been the winning card of a fleet in a world war, something that unfortunately seemed increasingly near and inevitable . . . . The atmosphere in Trieste was wonderful: poetry was recited in cafés, unpublished stories were read, there was talk of science and travel. It was June 1899. News of the cartographer and astronomist, Malcolm Fergusson, a member of Moore's scientific expedition, came from Africa. They discovered that Mt. Mfumbiro did not exist but was an invention of some charlatan's imagination. You know, this was important for my stories. I had always taken Mt. Mfumbiro as the metaphor for a peak that should not be crossed. You know, Giovanni Pascoli: "Oh mountains! oh rivers! it was better to stay, to dream, not to look beyond . . . . Dream is the infinite shadow of Truth."

"Uncle!" I exclaimed in exasperation.

"Oh yes, the paint. The tissue paper sent by Imelda in the box of candied fruit explains everything. Read it carefully. There are even the Kekulé formulae, the chromophore group, the chemical structure and your thermodynamic nonsense. You should read Goethe's *Theory of Colours* – it leaves Newton for dead."

Time passed slowly in the old tufo house. A drop of port from the Bohemian crystal decanter in the cherry cabinet. An oval Xanthia or a Turmac of salty Turkish tobacco. Thin and pale, his long hair now grey, uncle Vieri showed me a wooden model of a ship that he had built. It was a Russian warship that had docked that year in the port of Odessa. He had mastered the art of model building in his summer holidays as an adolescent at Forte dei Marmi. The school was that of my great-grandfather, Vieri's grandfather. For one who had made models of brigantines, complete with masts and sails, that model of the Imperial Russian Navy battleship had been child's play. In any case, the *Potemkin* found its place among the other ships on top of the bookshelves.

"Take Imelda's note and make good use of it in Lugano. Good luck for your conference. Remember to take Imelda the fig jam. She'll look you up during the meeting. Wait! Take her two fossils from my collection: an oyster and a small scallop from the Siena Pliocene clays. And don't forget to give her a hug from me . . . and *do* be careful: I understand the Aprilia can do 125 km/h."

As I descended the steps I thought of the love story in Odessa that had left traces in many of Vieri's stories: a sensual affair in the short summer of 1899, a love lost in the mists of the Black Sea and in the complex events involving Odessa six years later, a love melting in the dark eyes of a Russian girl in an old brown photograph. She had been a pretty young nanny, perhaps of Armenian origin, maybe from Samarkand, whom he met on the steps of the port of Odessa, as she wheeled a baby in a pram: a dark-eyed girl, impossible to find, impossible to forget; fresco of a closing century, with its spices and embroideries, its cafés and daguerreotypes, like Mfumbiro, the mountain that was not found.

At the Joint Congress of European Chemical Societies held in Lugano in autumn 1938, the Clausius medal was awarded to a young Italian scientist. The title of his communication was *Ferromagnetic and ultrasound studies on underwater paint: the production of an old Triestine factory revisited*. Thus, the secret research of German scientists on ultrasound and submarine paint for U-boats became Pulcinella's secret, and obviously it was not a case of exporting military secrets into Switzerland. The German scientists called it an unlucky coincidence.

Needless to say, the young Italian scientist made a big show of his Clausius medal and his new Aprilia. Not only did the existence of an uncle Vieri at Rapolano Terme go unmentioned, but it seems that one evening the young Italian was bold enough to say that the secret beauty of his berline's enamel was due to certain special submarine paints. Many years later, the German Bruker Spectrospin used a wonderful yellow paint on some small submarines built for peaceful purposes and underwater research.

The day after the congress, a smaller and more human Italian scientist was seen on the lake promenade, arm in arm with a fascinating elderly lady who worked at the Italian embassy in Switzerland. Her violet chiffon dress was caught by a breath of wind, revealing a pair of still beautiful legs, as her slender hand posted a card with two signatures and the usual warmest affectionate greetings. The message, in Cyrillic, mentioned a young dark-eyed nanny, a cruise to Odessa on the Black Sea, delicious evenings bathing and iridescent colours that can sometimes be seen underwater. It sounded like the verses of a famous old Russian love song. To Uncle Vieri, Rapolano Terme, Siena.

8

The mystery of Nemrut Dagi

The second day, Rosaluna brought Vieri a Burmese harp, a hundred years old, found by Veliero Edoardo Verdiano in Myanmar.

"The symbol of a film you enjoyed," she said. She was wearing a close-fitting T-shirt with a quotation across the chest: "I've seen things that you humans never dreamed: battle ships in flames off the bastions of Orion, beta rays flashing in the dark near the gates of Tannhauser and all these moments will be lost in time like tears in rain."

"I liked *Blade Runner*," replied Vieri, thanking her. "It is so different from the apprentice sorcerers of today, more human than cloned beings which a certain modern pseudo-science seeks to impose through unnatural genetic manipulation. Great, Rutger Hauer!"

That morning Vieri was reading Oscar Wilde, one of his favourite writers. The book was open at the page where Wilde defines a cynic as one who knows the price of everything but the value of nothing.

"Today's story is time-shifted too," said Vieri. "We go back to the 1920s. You know Pietro Cascella and Cordelia von den Steinen, don't you? We take them back many years. Again the story has nothing to do with Yucatan. Desire to reach a destination, curiosity, determination and the pinch of mystery that every adventure needs, these could all be useful for your journey."

A bet

The two friends generously paid the driver at Wadi Mousa, the river of Moses. The rest of the journey was on horseback. In 1923 that was the only way to reach the pink city of Petra.

The friends were both Acquarians born in early February, a dozen years apart, but stone was what brought them together. The older man saw stone as something to mould: his large sensitive hands imagined forms and ideas in stone. The chemistry graduate, on the other hand, was attracted by the composition of stones and their history; he sought out fossils rather than rare marbles and warm travertine. The travertine of Rapolano Terme was where their journey started. A few months earlier they had met in the old tufo house of the young graduate's uncle Vieri.

That evening, Vieri had been reading Verne's *Around the World in Eighty Days* in his velvet armchair with the Irish rug over his knees. They teased him about his literary preferences and accepted a glass of Nobile di Montepulciano. Among the smoke of Turkish tobacco, bets on the journey they were about to make and an inappropriate nostalgic Abruzzi shepherd's song on the radio, two crazy ideas emerged. The first was a Verne-inspired bet: Siena to Petra via Damasco, Jerusalem and the Negev desert in eight days. The second was a dream of the sculptor: to find the giant stone heads carved 2000 years ago and said to be on a mountain top on the eastern road, beyond the city of Edessa, beyond the furthest outposts of the crusaders in southeastern Turkey, where the brother of Goffredo di Buglione ruled.

"Edessa, now called Urfa," said Vieri, "was once a great civilisation. To the north you could see the Euphrates and Curdish territory. The Curds are a proud people who smart under Turkish domination and whose language is no longer taught in Turkish schools, but is handed down in poetry and epic poems. If these giant stone heads are somewhere, they could be at the summit of Nemrut Dagi,[1] above 2000 metres,

---

[1] The ruins of the temple of Antioch I of Commagene, also known as Filoromeo or Filelleno, on Nemrut (or Nimerud) Dagi, were only discovered in 1953. The burial chamber of Antioch I and its access corridor have remained a mystery. The stone statues, about nine metres high, represent seated figures; their heads, still integral, lie on the ground beside them.

though it is a mystery what stone they used. Those Mesopotamian mountains are of crumbly consistency; they are certainly not rocky."

"But the stone could have been transported from a quarry by mule and horse," interrupted the sculptor. "For a great sculpture, the stone is important and the place to erect it must also be right. Nature may nevertheless have placed monoliths on that mountain by volcanic eruption or meteorites."

"The mysterious monoliths on the mountain indicate that the place was sacred. Meteorites seem more likely than a volcanic eruption. There is a volcano of the same name further east, overlooking Lake Van, and its crater at 3000 metres is now occupied by a lake," continued uncle Vieri. "The great heads must be part of a temple to Antioch I, son of Mithridate of Commagene, erected in the first century BC to commemorate his reign, or as his tomb. Its doors face in three cardinal directions. It must be a rare example of syncretism of Hellenistic, Persian and Hittite traditions, related to the cult and mysteries of Mithra. In the Zoroastrian tradition, the bull is killed by Mithra, as in a modern *corrida*. From its tail spring ears of wheat. By dying the cosmic bull gives life. And here Armenian, Babilonian, Etruscan and oriental traditions in general form a single cosmic garland having its ethical foundations in agrarian religion and the sacredness of nature. I'm convinced that Nemrut Dagi is the place. A Turkish geologist who saw ruins up there in 1881 told me about it. Unfortunately in 1899, during my journey to Lake Van, I was unable to find a way into the area, partly for lack of time and partly because the Curdish guerrillas were particularly active. However, during the trip I met a great German archaeologist, Otto Puchstein, renowned for the altar of Pergamon. He showed me a photo of a bas relief depicting Apollo and Antioch I who sported a typical Persian-Babylonian headdress: it came from Nemrut Dagi and the funerary monument of King Antioch. Today, after more than 20 years, that discovery is no longer mentioned. Puchstein is dead and took the secrets of the Commagene lineage with him."

Vieri drew his golden Eberhard pocket watch out of his gilet pocket: it was late. They bid each other farewell.

Petra

Vieri's information and advice had been precious for that crazy race to Petra: the right Lloyd Triestino liner, a book with detailed road maps and even more detailed notes, an old Hoepli manual and his friend the sultan of Aleppo who welcomed the two travellers to his princely residence, offering them fruit with ice, perfumed thermal baths and above all, an automobile with driver for their venture. Like the grape varieties of a great wine, the right combination of adventure, wisdom, instinct, logic and culture made Vieri an unusual personality at the turn of the century. We could describe him as a fine detective or a refined author. His travel stories and intuitions were always a magic combination of science and art, ingredients of the friendship that bound our two rambling men.

The bet could be considered won. Seven and a half days since they left Siena, our friends were strolling in the narrow gully of Petra between churches carved in the rock face, tombs and imposing monuments. They were discussing the role of form in art and science, and the beauty of El-Khazneh and the Great High Place, the principal monuments of that abandoned city. They remarked that three years had passed without anything being published on the extraordinary archaeological finding of Miss Conway in 1920. However, their minds were already with the Curds and among the heads of Antioch at the summit of an imaginary mountain in eastern Turkey.

On their unhurried return, a stop at Aleppo was not so much an obligation as a longed-for excuse for idleness; it lasted many weeks. The old friend of uncle Vieri had moved into the summer palace where azure and green peacocks strolled among Lebanese cedars and *Olea fragrans*. One evening at dinner, the Arab prince announced two surprises. The first was a colourful hot air balloon for the ascent of Nemrut Dagi, together with an escort of carts, porters, food and whatever an expedition of Marco Polo might need. The second was a guest. The young German-speaking artist stole the imagination of the sculptor. His eyes were captured by her glance and her figure. Cordelia's high-sounding name, von den Steinen, meaning "of the stones," also contributed to her aura. She was destined to radically change the life of the Italian sculptor.

After six courses and many libations, discussion became animated, as between old friends. The prince, well read in astronomy and alchemy, raised two intriguing concepts: the role of time in art and science, with special reference to the creation of forms, and the importance of the natural environment that factories and coal had begun to destroy, especially in northern Europe. The elderly Arab mentioned the controversial young English writer, David Herbert Lawrence, whose recent novel, *The White Peacock*, he had just read. He had been attracted by the title (he was very fond of peacocks) as well as by the book's *ecological* content, *ecology* being a term coined by Ernst Haeckel in 1866 for the science of the natural environment that he was studying. He remarked on the English writer's interest in Etruscan culture and his intention to go to Italy to study Etruscan places, retracing the steps of a people much more oriental, wise and complex than the Romans with their wars, roads and cold marble statues.

"I couldn't agree more!" exclaimed the sculptor. "Time and stone are the point!" Glancing at Cordelia, he added, "Etruscan culture is the fruit of systemic and cosmic knowledge, so well represented in their humorous smiles. A culture that found the path of complexity in the cycles of nature, in the colours of the earth, in rejection of dualism. Thus, death was a source of life, creativity, art and ecological balance. The deer devoured by the lion: deers were spotted for Etruscans, meaning day and night. Lions were dark and pale: death and life are in essence the same thing."

Silver platters of exquisite grapes and fragrant figs were brought to the table. The young scientist spoke of the arrow of time. The young German artist was excited to hear discussion of the role of time. A few years earlier in Freiburg she had had a long and passionate relationship, perhaps a love story, with a Viennese physicist. His name was Erwin Schrödinger.[2]

---

[2] Ten years later, in 1933, Schrödinger shared the Nobel Prize with Paul Dirac and began work on his great book "What is life?" where he introduces the concept opposite to entropy – disorder, namely negentropy – organisation – complexity, in other words *life*.

"Erwin," she said, "often talked to my mother; not so much about his discoveries of quantum mechanics, but about a new idea that he linked to time and form: *negentropy*. During walks in the forest, he used to say that living beings feed on negentropy, the opposite of disorder and pollution which are measured by *entropy*."

"I too am concerned with the role of time in physical chemistry," added the young scientist. "I think it can be linked to the concept of form. There is an aesthetic of nature evident in the structure of molecules, in the designs of shells and butterflies, in the colours of flowers and feathers. Take peacocks, for example! Biological evolution is evolution of forms and structures, modelled by time as if by a sculptor's hand. Stones, with their history, are impregnated with time. Time is in matter. Time is in the history of this beautiful planet, the blue orange on which we have lived for millions of years."

"A fine image, that of the blue orange," said the sculptor. "An oxymoron of rare beauty. If I had time and stone I would make a cosmogony, the cosmogony of the blue orange!"

The hot air balloon

Next morning, the three young people affectionately bade farewell to the old prince, promising to return to Aleppo within three years. The journey towards Kahta, the village where they would start for Nemrut Dagi, was neither comfortable nor easy. Despite the princely escort and transport worthy of a pharoah, problems arose every day. As soon as they crossed from Syria into Turkey, they realised that the political situation in Turkey was precipitating. The People's Party had found a natural leader in Gãzhi Mustafà Kemãl and evolution towards a modern form of power was proceeding at a fast pace. Ethnic and religious divisions were exasperated, with Curds, Muslims, Orthodox Christians and Armenians finding pretexts for bloodshed. The Curdish area of Nemrut Dagi was of course among the hottest.

Despite the difficulties, in the afternoon of 23 October 1923,[3] a fine colourful balloon could be seen over the mountains north of Kahta. In the basket, next to the red jersey of a young Italian and the green Abruzzi wool sweater of a big man with large hands, the violet shirt of a German girl stood out. It was nearly sunset as the balloon laboured against opposing winds to attain 2000 metres near the mountain top. Suddenly, as the sculptor was pointing out the large stone forms on the top of the mountain, the sound of approaching planes was heard. Two war biplanes were within firing distance of the balloon in less than no time.

A month later, different versions of the event were reported on the *Campo del Sole*, a liner carrying Cordelia and her two friends from Smirne to Venice. What insignia did the biplanes carry and why did they open fire? This question remained unanswered. How was the balloon saved? Cordelia and the young scientist both claimed credit for bringing the balloon down among heavily wooded hilltops where the assailants could not see them. However, the versions diverged as regards the sculptures of Nemrut Dagi, which remain veiled in mystery. Cordelia and the scientist did not see them, while the sculptor claimed that the massive head with the Persian-Babilonian headdress wore an enigmatic smile. It seemed to beckon them through the rays of the setting sun. It must have been nearly three metres tall.

The bet was paid. Uncle Vieri gave his nephew three beautiful collector's pieces of travertine bearing the imprints of leaves, fish and insects, as well as a jewel received from one of his many friends throughout the world, a Creole lawyer from Santo Domingo. It was a piece of perfectly transparent amber, about the size of an egg, enclosing a tiny perfectly conserved orchid.

The sculptor was invited to go to the window of the old tufo house. Below in the square were eleven trucks parked in a line, each bearing a

---

[3] Six days later, on 29 October 1923, Mustafà Kemāl was elected president. He later took the name Kemāl Atatürk.

block of travertine about three metres high. The blocks were from different quarries in the Rapolano area, each of a different hue.

The Diatto and the peacock of Aleppo

The following spring, grandfather Edoardo Quirino offered to accompany his nephew, the scientist, and uncle Vieri to the Lunigiana in his antiquated Diatto. The car's number plate was SI-10 – he had purchased it in 1909. It was a large chunky white car with bellows like those of carriages.

After hours and hours of dusty road they reached the Verrucola of Fivizzano where the sculptor lived. From the castle portal, a girl in a violet blouse ran to meet them, carrying a handful of fragrant prugnoli, the delicious spring mushrooms of Lunigiana that are eaten raw.

In the depths of the castle they found the 11 blocks of travertine, transformed into spheres, spirals, arches, stars and horoscope tables. Deers, lions, eagles, leopards, giant bees, Etruscan chimeras, triangles and Persian-Babilonian headdresses were taking form. The main piece was a balancing sphere of time and the columns of Ulysses, a metaphor for biological evolution and the unsurpassable limits of the blue orange, our planet: a syncretic cosmogony of the natural history of the Earth. They seemed to detect the perfume of incense of imaginary altars, but there was no smoke. Magic and Etruscan smiles had entered the great halls of the castle through art. By now it was dusk, and they saw a great porcupine (*Hystrix*) emerging from the nearby wood.

The large handed sculptor promised to transport all the pieces to Nemrut Dagi. He recounted the ancient Arabian cosmogony of the sun and moon who invited water to their house and had to climb onto the roof and then into the sky. He told them about Doondari, the Massai goddess, who came to Earth and created stone. He told about the age when it was possible to touch the sky with a hand, a legend handed down by the gypsy Shim.

These reminiscences and tales would have gone on, had not the peacock launched its raucous cry from the cage mounted on the back of the Diatto. The friends ran to open the cage and the proud peacock of Aleppo flew out and settled on the top branch of a nearby olive. In distant Rapolano Terme, its brother perched in another olive tree near the tufo mansion, from which the friends had departed a year before to solve the unsolvable.

The gifts from the Arab prince were accompanied by cards: one bore an inscription about the science of time, the other about art and love, quoting verses from the fourth *sura*. The names of the recipients were written in Arabic, but it was not necessary to decipher the signs in order to understand for whom they were intended.

9

Papier-mâché soldiers

For Your Five Senses

Your skin is nourished
by the sun warmth
your skin keeps on
the strong old smell
and your cheeks are imbued
by the sweet bitter amber
my lovely young girl
perfuming of orange blossoms.

Time doesn't stop
in the cutting glares,
time runs following dreams
with running nistagmi
that are black, fast and full
of sadness and love
my lovely young girl
with the colours of orange blossoms.

The old Goethe dreamt
fixed and strange moments
but the echo came back
from the aider forest
but the echo played back
loud and soon far,
my lovely young girl,
its African dance.

I'd like to feel again
in the old kitchen
tuscan majolicas
in the wooden cupboard,
the taste of the honey
dripping down from the looms
my pale young girl
running to no-time.

I'd like to touch you strongly
I'd like to touch you softly,
caressing your skin
with my hand palm
older tenderness,
present tenderness
my lovely young girls
with smiling visages.

On the third day, Rosaluna brought the third gift. It was a black Etruscan cup that one of uncle Vieri's brothers had found on an official archaeological dig that had filled the museums of Chiusi, Viterbo and Siena with exhibits.

Before the conversation turned to the apprentice sorcerers of genetic engineering, Vieri proposed a reflection on several films. The story that day would be unfinished, leaving room for the reader's imagination, like scenes for a film. In some ways, it would be a continuation of the previous day's conversation on *Blade Runner*. He asked Rosaluna's opinion on *Nirvana* directed by Salvatores, *Matrix* starring Keanu Reeves and Akira Kurosawa's *Dreams*. Rosaluna considered them excellent. She liked the scene of the peach trees in *Dreams*: a poem to nature, to spring, to the imagination, in which dolls left their showcase and became real. The colours of their costumes weaved among the pink blossoms on the grassy terraces of the orchard. According to Vieri, the dolls that came to life were like the papier-mâché figures of nativity scenes from his childhood.

Nativity scene with Indians

Father Raimondo Bernini, priest of San Domenico church and "correttore" of the Drago contrada of Siena, invited the panel of judges of the nativity scene competition to see one created by the Siena branch of Vieri's family. The scene, mounted on boards that rested on wooden horses, occupied a whole room in a fine house in the centre of Siena, with views on the Cathedral, San Domenico and the Torre del Mangia. The house no longer belonged to the family. On that sunny day in late December, a tiny red-tailed bird flew in the open window and pecked among the statues and fresh moss spread on the boards. Melchior, approaching Bethlehem with his two companions, was overturned.

The judges discussed the scene with animation and finally awarded it first prize, partly due to the representations of father Bernini. It was, indeed, a special nativity scene. Divided in two parts, the second "unorthodox" part, beyond some papier-mâché mountains, sported an Indian camp on the banks of a silver river, with the wigwams, totems and mustangs of a happy people who had not yet heard the "good news." The papier-mâché Indians were by the famous craftsman Carlo Confalonieri of Milan, tiny works of art. The children of the family would move them around, imagining battles described in Emilio Salgari's *Le Selve Ardenti*. They had named two of the squaws Yalla and Minnehaha after characters in the book.

"Perhaps the Wise Man with the dark skin reached Bethlehem from the Indian village," surmised one of the children. "Let's hope he brings more papier-mâché figures for Epiphany!"

White nights, swallows, lichens, reindeer, hovercraft and quicksand: stories of freedom

A herd of thousands of reindeer, raised in the wild, was not something one saw every day. In June 1900, the reunion of Sami at Lake Kilpisjarvi, about 300 km north of the Arctic Circle, where today the borders of Finland, Norway and Sweden meet, was a special event. The occasion was the beginning of the new century. The three visitors who were there by chance for completely different reasons considered themselves fortunate to be present.

Fine red dust tinged the *kota*, the tents of reindeer hide, the birches and the red and blue costumes of the Sami, gnomes of the great Arctic tundra. The shamans smoked their pipes and sold horn amulets and reindeer-horn fertility powder. Girls with exquisite ancient gold pins on their breasts offered iridescent reddish-amber cloudberries, gathered on the heath.

Sir Wilfred Mordred, one of the three foreigners present, was not interested in reindeer or cloudberries. Ornithologist from Cambridge University, he had ventured into the Arctic to study the behaviour of the Lapland bunting and the great grey owl, *Strix nebulosa,* with its yellow concentric-circled eyes. He needed a sensational discovery to distract him from the state of his coffers, which had been depleted by sprees at the dog races. He was counting on interesting data on the nesting habits of the little rusty-necked bunting, previously reported by him as an occasional visitor to Scotland and Ireland, or on the nocturnal behaviour of the owl in the month of the midnight sun, when the night, or rather darkness, does not exist.

The erudite dissertations of the English professor had clearly bored the slim blond Italian with the Tuscan accent. While sharing Mordred's passion for ornithology, the 25-year-old was much more interested in the busy comings and goings of house martins tending their nests under the eaves of the wooden houses near the lake. Though the mosquitoes were annoying, they provided plenty of food for the birds, which would occasionally pose on the ground in the long bright nights. The white-backed house martins were more sociable than the Italian ones and slightly smaller. These Arctic sisters were still busy foraging at midnight and went to roost at dawn, namely a few minutes after midnight. Three or four little heads peeped out of each cosy mud nest. House martins nested in Lapland from May to September.

Satisfied with his observations, the Italian rubbed his hands together at length, a habit of his when he felt at peace with himself and with nature. Two months earlier, Lord Brassey of Brassey Mines Ltd. had charged our subject, Zeffiro Maria Vieri, with a mission in Kiruna, Sweden, a mining town near the border with Norway and Finland. The township had been established that very year in the site of a Lapp settlement. Kiruna was said to have the richest iron deposits in the world, and

Lord Brassey was worried about Russian (Finland was under Russian dominion) and Prussian expansionism. The British company had chosen an Italian in order to be unobtrusive. His job would be to collect data on the potential of Kirunavaara, the big magnetite mine, and to get the Kiruna–Narvik railway built in order to ship out the ore. The job was merited, after Zeffiro's role in another business involving the ancient mines of blende and galena at Ingurtosu, the sinkhole of western Sardinia.

The English entrepreneur had visited Zeffiro's father in the old tufo mansion in Rapolano Terme, Siena. He had known the Tuscan family for some time. The small hardware firm and the mining company had done business for a number of generations. On that particular occasion, a barrique of vinsanto was opened in the attic and the cherry sideboard acquired a new Bohemian crystal decanter: vinsanto vintage 1875, the year in which Zeffiro Maria Vieri was born.

Once in Scandinavia, Zeffiro had not missed the chance to visit Lapland for the annual gathering of the reindeer herds and to experience the midnight sun. He was in a holiday mood, after the work done in Kiruna and Narvik. The area of Lapland between Sweden and Norway was incredible. Kiruna Province was vast: 20,000 square kilometres, an endless plain of beeches surrounded by high mountains. The road from Kiruna to Narvik, which would be the route of the future railway, flanked a lake for 70 kilometres. Part of the area was of great natural interest, populated by bears, lynx, moose and lemmings. In the long white nights, Arctic hares could be seen racing: surreal in a surreal landscape. For eight months of the year, the flat solitude of the birches was blanketed in snow, for the other four it was swathed in white light.

At the start of the 20th century, a pioneering atmosphere typical of mining towns pervaded Kiruna, and a sort of artistic avant-garde. Driven by the potential of 20 million tons per year of iron, the township was growing. It would have a wooden church in the form of a Lapp tent, the famous Hotel Ferrum with its casino, a 23-bell carillon and strange artistic creations devised by bizarre modern architects. It would share its surreal fascination with Alvar Aalto's Rovaniemi in Finnish Lapland, like the summer sun that sets, only to rise again nearby a few minutes later.

Zeffiro got away as soon as possible and headed for Kilpisjarvi and its reindeer. He had money to spend, won at the casino betting on 27, which had come up four times in a row.

The third visitor was not exactly a foreigner: a pretty blond girl with a husky voice from a *torppa* or farm in Finnish Karelia. During the long white nights, she sketched cloudberries and the amazing forms of pink lichens, seeking the fretwork and lacy profiles of natural surfaces. Today we would say she drew Mandelbrot fractals.

During frequent walks on the heath and trips on the lake, Zeffiro and Aila became friends, discovering many ideas in common. They often spoke of the relation between science and nature. Zeffiro had studied agriculture at Florence University and was excited by the young Finn's criticism of mechanistic descriptions of biological species. According to Aila, Euclidean geometry was not appropriate for describing nature. A science that did not envisage quality, evolution of forms, relations between components of an evolving biological system, could not provide scientific answers to the complexity of the natural world.

Many years later, long after the tragic events of those days, comfortably seated in the crimson velvet armchairs of the tufo mansion at Rapolano Terme, Zeffiro realised that certain of Aila's drawings contained premonitions of *art nouveau*. He discovered that Aila was of the school that founded Finnish industrial design which had great success in craftsmanship and furnishings, leading to the artistic notoriety of Alvar Aalto and Aino, wife of the Finnish architect and producer of Artek curved-wood furniture.

Three days after the Sami gathering, two unsavoury characters arrived in Kilpisjarvi. They were wearing long trench coats and dark glasses, and they spoke Russian. The Finnish girl became visibly nervous and disappeared for several hours, taking the road south in the direction of Muotkatakka. Next day, she asked Zeffiro to come with her to gather bilberries and cloudberries on the heath. She asked him openly in Finnish. The Russians were sitting at the next table.

That evening, Zeffiro returned late in a pitiful state. He announced that he had lost sight of Aila. The last time he had seen her, she was in the distance, towards the tundra that leads to Muotkatakka. The heath seemed to have swallowed her.

"The quicksand," cried Sir Wilfred, losing his usual english aplomb. "The area is full of quicksand, it is almost impossible to avoid it!"

Tension was high in the Lapp village. The silence was broken only by the distant roar of the seaplane taking off from the lake. The search continued throughout the white night but was unsuccessful.

The Russians introduced themselves officially as special police agents of the tsar Nikolas II, a cross between bureaucrats and military police of the worst species. The previous year the tsar had abolished the Diet, cancelled the constitution of the Grand Duchy of Finland and repressed the independentist movement in Karelia. The agents asked many questions about the girl and demanded a detailed account of the day from Zeffiro. His Italian passport saved him from further complications.

Sir Wilfred Mordred did not spare them his irony: "The girl was obviously in a spot. She took the solution of the lemmings, that here in Scandinavia commit suicide by leaping off cliffs. The only difference is that she opted for quicksand."

In the summer of 1901, just after the Palio of 2 July, a year after this unfortunate event, the Caffè Greco in Siena was buzzing with customers. At a table in a secluded corner, Zeffiro Maria Vieri was opening two gifts from a pretty blond Finnish girl who had chosen freedom in a flying boat the previous year in Lapland. With her, she had taken the secrets of the independentist movement of Karelia. One of the gifts was a drawing of two swallows peeping from a mud nest. The other was a precious, blown-glass vial bound in gold, containing reindeer-horn powder that Aila had obtained from a great animist shaman of the Sami tribe of Mt. Saana.

Much to Lord Brassey's satisfaction, the Kiruna–Narvik railway, owned by a new Anglo-Swedish company, was inaugurated in 1902. An enterprising young Italian was said to have played an important role.

Le secche di gennaio

"You will love my coloured glass lampshades," she said.

Zeffiro Maria Vieri's mind was elsewhere, as usual. He was following the migration of some small flocks of finches. "They are gregarious," he replied. "At the *secche di gennaio* (the January dry period), droves of them come through Sardinia . . . you were saying?"

"I was thinking of furnishings for the house in Alghero. Tomorrow the trunks arrive with the boat from Leghorn: the books, double-coloured panes for the doors and verandah to defend us from the *mistral*, your radio . . . you couldn't survive two months without the Imca! I was saying that you will love the lampshades, some in ground glass, others coloured with Klimt figures, others with beads." Aila's eyes betrayed a mixture of reverie and fondness. A second flock of finches flew low over their heads towards the sea.

"See! You can recognise them by their flight. They bounce along."

The air was salty on that mid-winter's day of 1919. The days had already lengthened perceptibly which heartened them both. Zeffiro said that the seasons needed oxygen to return and that oxygen comes from the deep breath of light which grows longer and longer until its apotheosis in June. Then the lungs of the Earth have to rest.

The *secche di gennaio* brought the mimosa into bloom. Already in mid-January there was an explosion of yellow. Aila and Zeffiro picked some sprigs and their nostrils grew yellow as they inhaled the scent. They ran hand in hand to the tandem bicycle leaning against the Mediterranean shrubbery beside the dusty road. They mounted and pedalled slowly towards Alghero.

Maurizio B. took leave of his friends earlier than usual. Next day he was embarking for Sardinia from Leghorn. He had already loaded

Zeffiro's trunks into the old truck. He read and reread the telegram from Ernst Zinner in Karlsruhe: the job would not be simple or without its problems.

The Wild Goose at Round Stone, on the Connemara cliffs, served the best prawns on the Atlantic coast of Ireland. The young American was sitting near the open fire in the company of an auburn-haired, green-eyed Irish girl. He knew her as Miss Entropy and did not dare ask for other details. As a student of physics at Washington University, he was well aware that it was a pseudonym, but he also knew that those were hard times for the Irish involved in the independentist movement and that the young biologist played a key role in liaison with Irish emigrants to America. Besides, her company by day, and especially by night, was very satisfying. It was a rare thing to make love and also talk about science at a high level. And *he* had chosen to be involved in that adventure. He had insisted on a different holiday with Zeffiro Maria Vieri, his Italian friend, and with his Austrian colleague, Ernst Zinner.

The young American was very tall and his varsity basketball team had reached the finals thanks to his rapid eclectic game. His name was Felix Bloch and he was at the start of a scientific career that would to bring him the Nobel Prize in Physics and lead to the discovery of the "Bloch equations," which would be a foundation of the theory of nuclear magnetic resonance. His research on cryomagnets and liquid helium, more than 200 degrees below zero, to produce very high electromagnetic fields, was in Zinner's view essential for the working of the Bruker Spectroscopein minisubmarine that would be a player in other adventures.

Many years later, the congress of the Italian Society for the Advancement of Science would award Bloch with a gold medal in a ceremony held in Sardinia, and the distinguished elderly American would have difficulty climbing into a red Fiat cinquecento that would take him round Sardinia to see the places he knew from 40 years earlier, but would have to pretend he didn't know, because the stay in Sardinia with Miss Entropy in 1919 had to remain an absolute secret for the rest of his life.

The friends decided to name the little yellow submarine after a Mediterranean shell, inspired of course by reading Jules Verne. Zeffiro proposed *Dentalium* or *Epitonium*.

Zeffiro sipped his ratafia, lit a Xanthia and began to think through smoke rings: Lord Brassey was impeccably English and would not sustain the Irish movement for any reason; but he was also a dear friend and could not be deceived. Thus, the operation could not start from the jetty of Is Piscinas at Brassey's mine, but neither from much further away, due to the limited range of the *Epitonium*.

He went upstairs to examine the charts that engineer Caroli had given him. There was no sufficiently safe and secluded dock along that coast, either north towards Alghero or south towards Carloforte. The *mistral* would make it impossible to prepare the little submarine even in fairly sheltered bays. The operation could only be conducted between the end of February and the first days of March.

That part of the Sardinian coast was suitable because the *secche di gennaio* are followed by the *maestrale* at winter's end. It was an insolvable problem: two equations and three unknowns, as father Baldi would have said. His mathematics lecturer at university would also have added that relations are what count, not things in themselves. Without relations, problems cannot be solved.

"I'll go and ask father Baldi," thought Zeffiro aloud. "I have owed him a visit since I arrived in Sardinia, now that he had retired to the parish of San Pietro in Bosa."

"San Pietro! Why didn't I think of it before? Enter the estuary of the Temo and sail a few kilometres up to San Pietro. It is navigable for small boats. Boats reach the tannery at Sas Conzas. From there it's less than two kilometres."

The *canonica* chosen by father Baldi to study and write, or rather rewrite, the essay on time on which he had been working for 30 years, was the ideal place for such an endeavour. A pinch of antiquity – actually more than a pinch, a generous dose: the small church had been built in 1073 using trachyte in various tones of pink. Its sturdy square-based bell tower, added in the 12th century, had two delightfully asymmetric windows in the upper part. A pinch of nature, with a rich orange grove that sloped down to the river. A pinch of cultural fraternity in that second floor loggiato with pink trachyte columns, the great olivewood table and the view of the distant castle of Serravalle built by the Malaspina of Genoa.

The five friends were seated with father Baldi, who had quickly cleared the olivewood table of leaves of paper and piles of books. The girls, Aila and Miss Entropy, had immediately found something in common, united by their commitment for the liberation of their respective peoples. A well-fleshed woman busied herself between the kitchen and the loggia with a bottle of S'Eleme, some delicious slices of mullet roe and *carasau*. Felix Bloch and father Baldi were already engaged in a noisy discussion in English on the lives of spins in quantum levels. "Trottole!" exclaimed father Baldi. He did not like anglicisms and indeed *trottole* are spinning tops.

"The Irish flying boat cannot land on Lake Omodeo as planned," interrupted Miss Entropy. "The Lake does not yet exist. Damming of the Tirso is years behind schedule. An alternative would be Cabras lagoon and northward on foot through Santa Caterina di Pittinurri and Bosa."

"At the same hour, the submarine could be anchored off the Temo estuary," said Ernst Zinner. Zeffiro nodded.

At first sight, Ernst Zinner would never have been taken for a scientist, a physics lecturer in a prestigious English university. Hands of an Austrian woodcutter, gait a cross between that of a bear and a stiltbird, rough spoken with an incomprehensible German accent, he was usually taken for an Austrian peasant. However, that primitive exterior bore two pale blue eyes that expressed incredible vitality and intelligence. In Ernst, scientific genius and tender artistic sensitivity found perfect harmony. Even his crude way of speaking became gentle when he described the colours and images of Gustav Klimt and when he discussed matrices and correlation times of magnetic interactions with Bloch. His natural artistic temperament was expressed through photography. He used a colossal mahogany daguerreotype camera and spent whole nights in the dark room, producing photos of great sensitivity or with alchemical tricks. Once he showed ghosts passing through walls and doors; in a print given to friends from Rapolano Terme, the elderly curate levitated above the faithful in his parish church. His skill was useful for the Sardinian operation; indeed, on that occasion he made a masterpiece . . .

This story began at the *secche di gennaio* and like all biological stories that move in cycles, ends at the *secche di gennaio*, that naturally return each year. They return in a slightly different way because biological cycles are actually spirals that evolve and never return to the same place at the same moment. Places, too, change with time and if one returns they are never as they were. This is the fascination and mystery of life, and the reason for uncertainty and fear. The same is true of this story, the ending of which is unknown. It can only be said that it goes in a somewhat random manner, depending on the choices of the players at appropriate times: like the street masters in the cartoons of Moebius.

Our story runs aground on the *secche di gennaio* and does not reveal any more than a narration permits: the facts, consummation of the facts, an ending, strong determinate events would not be in line with the *secche di gennaio*. The story described the atmospheres, the outlines, the moods, the places and some characters, but it did not reveal the plot or consummate events. It is up to the reader to use his/her imagination and find conclusions of different kinds. The secrets of events remain hidden in the *secche di gennaio*, also because the characters would not have wanted them to be told.

Several years later, Miss Entropy was seen in Ireland. It is well known that Felix Bloch was awarded the Nobel Prize for Physics, less known that Ernst Zinner recently received the Blaise Pascal Medal of the European Academy of Sciences.

Before the great Etruscan fire place in black basalt, Zeffiro Maria Vieri tells all the stories he was part of between the end of the 19th and the first half of the 20th century. He never hesitates to answer questions, whether about secrets or the endings of stories, but do not ask him about the *secche di gennaio* because in that case he always replies that the sacredness of time and nature must be respected and that even the most experienced sailors run aground on the *secche di gennaio*.

Milongas and armadillos (part of a story told by Edoardo Quirino)

The steamer of the Navigazione Generale Italiana, *Giulio Cesare*, left Genoa on 25 October 1922. The graduate thesis on rare earths of one of Edoardo Quirino's nephews could wait another year. The trip was not to be missed. Edoardo Quirino, now almost 50, had decided to interrupt his inborn laziness and do some of the things he had done in his youth. He had invited his nephew to go with him. Their destinations were Argentina and Uruguay, where they would visit a pure-blooded Uruguayan, meaning at least 51% Spanish and a combination of indigenous, Carib, Gypsy and Italian.

"Bel Romeo," as Quirino called his South American friend, y García y des Olivos y Ross y Osceola y Reynaga y Gamusa y Casa Baclò y Pedrera y Armendáriz y de Viñaroz, was a long-distance sailor of few ships, in his own words ship's captain, in Quirino's words winner of an unexpected fortune in sterling gold. Since he had become rich, he had bought a large, two-storey wooden house with attic, on the ocean near Punta del Este, surrounded by a thousand hectares of park, animals, gardens, water and forest.

There were four versions of how he came upon the unexpected fortune. The first was that he inherited it from the Carib side of his family, an old pirate from Belize, retired from business; the second was a lottery won in Italy or a lucky night in one of the casinos of the Côte d'Azur. The third was related to the gypsy side of the family: part of the hereditary treasures handed down by gypsy kings at the Camargue gathering in 1912 and the testament of Romeo's great-grandmother, a gypsy queen. The fourth involved a rich Irish lady, fascinating though elderly, who was said to have lost her head for him and left him everything. Edoardo Quirino, in a serious vein, had stated that the second and third hypotheses could be excluded because the sailor had restricted the versions to the first and fourth, one night after a bottle of Calvados. "Knowing Romeo, a fortunate combination of both cannot be excluded," added Quirino. Indeed, the beautiful green-eyed, copper-headed Siàn O'Harp had died at the tender age of 69 and in 1912 Romeo had sailed, at least for once in his life, with his Indian Creole friend Gamusa senior to the Belize keys and Terranova.

Romeo Ran y Gamusa, as appeared more simply on the gate of his villa at Punta del Este, was born on 10 July 1887 and had reached an age in which adventures were still strongly desired, though with a note of saturation, tempered by memories, knowledge of many peoples and countries, love affairs, bottles of port, a pinch of acquired wisdom and of course the idleness induced by a beautiful house on the Atlantic and the company of a splendid young lady called Turquoise Primavera.

Actually, Romeo's exact birthday was a mystery that even his most intimate friends were barred from asking about. Romeo would reply, "If you want to know more, ask signora Bocca Dorata of Queimada Island or that cartoonist of yours, from Venice I think, the cabala expert." This was his constant answer, and nobody until then (1922) had been able to understand what he meant or who the two characters were. In any case, partly out of faithfulness to the mystery and partly so as not to count the passing years, Romeo had a golden rule: he did not have birthdays.

Thus, Edoardo Quirino described Romeo to his nephew on the voyage from Genoa to Barcelona, in the private double apartment, all mirrors and brocade, reserved for them by Romeo aboard the *Giulio Cesare,* or in the first-class bar and smoking room, where they relaxed on a small mahogany divan near a larboard porthole. The description was rather summary, considering Romeo was a man of many adventures and many secrets.

Edoardo Quirino had been convinced to undertake the voyage not so much by the invitation, as by money for the trip sent through the Buenos Aires office of the Istituto Italiano di Credito Marittimo. This had overcome not only his laziness but also his thrift, a family trait sometimes broken by sudden excesses of generosity. It should be said, however, that in every circumstance, Edoardo Quirino was and behaved like a perfect gentleman.

Evenings on the steamer were pleasurable, one reason being that they had found two bridge partners. They were two loners: an Emilian called Glennis, prematurely bald, had been seated next to the nephew during an afternoon projection of Fritz Lang's *Metropolis*. They immediately became friends, sharing tastes, enthusiasms and curiosities. The second was a conquest of Edoardo Quirino in the smoking room.

As he aged, the Tuscan gentleman had changed his tastes for women and liqueurs. He was now only attracted by plump women and only drank sweet liqueurs. In the cherry cabinet of his study in the old tufo mansion at Rapolano Terme, there was always a bottle of Strega and Meletti Anisette with the inevitable Bohemian crystal decanter containing the best port. Sometimes he produced a fine Maraschino di Zara or the black cherry ratafia of the prize-winning house of Adorno. Edoardo Quirino had succeeded in convincing the attractive, shapely Polish passenger in the next cabin, Mme. Maria Eva Kozlawski, Mariu for friends, hat stylist, but actually a little bit poet, a little bit artist, a little bit involved in middle-European avant-garde cultural movements, of two important things. The first was that a lady of great taste could not continue to drink vulgar spirits but must taste the divine ratafia of his personal reserve. Mariu condescended and fell in love with ratafia. The second important thing was that the best bid in bridge was not the antiquated English method but a complicated conventional system recently developed by four dear Neapolitan friends of Edoardo Quirino. Mariu learned the Neapolitan bid and became Quirino's inseparable bridge partner in the long nightly games against his nephew and Glennis. Apart from Edoardo Quirino's colourful comments on Mariu and her feline gait, sensuous communion between the two was limited to ratafia and the Neapolitan bid.

Maria Eva, with her quasi-erotic language, drove Glennis crazy, and he vindicated himself by distracting his adversaries with jokes, his cabaret numbers, his infectious laugh and by passing information to his partner through winks, signals and inadmissible miming. For Edoardo

Quirino, it was a heresy that Glennis continually ordered vulgar grappas, and sipped them in the manner of a connoisseur, guessing their names and origins. Not to mention his horrible American cigarettes, Lucky Strike, with their detestable, ill-boding, sickly smell that clashed with his refined Xanthias and Turmacs! The result was good bridge, lively conversations and not a trace of boredom.

Glennis was a "creative," as he liked to call himself. In his studio in Reggio Emilia he drew and invented images that became the advertisements and covers of *Arancia Blu*. He proudly showed them his latest creations from *Le Vie d'Italia*, the monthly magazine of the Touring Club Italiano: a flash of light for lightning gasoline, the electric tenor or modern automobile horn, and a crown consisting of the jewels of motorism (Diatto, Scat, Fiat, Itala etc.) for Oleoblitz, the Italian lubricant.

The interrupted story of Edoardo Quirino included some notes written in ink about an excursion in the pampas and an armadillo that ran from its burrow, not to mention Carlos Gardel, tangos and milongas. The sheets showed a sketch of the Valdez peninsula in Patagonia, and Edoardo Quirino told them about the time he touched a whale and her calf from Romeo's boat. Glennis seized the idea of the whale and drew a blue cover for *Arancia Blu*.

Absinthe dream

(Rosaluna found this story that uncle Vieri had hidden in his walnut writing desk)

I can resist anything but temptation.
Oscar Wilde

The sun had just risen and shone pink on Cima Dodici. Two female mallards breasted the current of the Brenta in Valsugana. The pink was different from that of the Dolomites, but spires of smoke rising from several points of the valley around Roncegno made the landscape worthy of Cézanne.

Unusually for Vieri, he had risen early and from room 118 of the Palace Grand Hotel, he gazed on Mt. Ortigara through the columns of the verandah where once the empress of Austria had sipped tea. Vieri was nervously smoking one of his blue oval Turmacs. It was unusual for him to smoke in the morning. He usually lit his first Turmac or Xanthia after his coffee at lunch. As a connoisseur of tobacco, he often followed his evening cigarette with a good Cuban cigar or a mellow Suerdieck from Reconcavo Bahiano, Brazil.

An absinthe dream from which he had just awoken had left him feeling nervous. Unusual for the phlegmatic Tuscan gentleman: the dream had been full of erotic references and he still felt shaken. He tried to imagine Rosaluna della Berardenga, as he called her, and had the feeling that she too was dreaming. He imagined her in the big bed at Pacina, at Castelnuovo Berardenga in the Siena Chianti. And indeed, at that very moment, Rosaluna was dreaming, not an absinthe but a honeysuckle dream. She was riding the hippogriff "pel sognato alone" into stories that led to other stories and opened turquoise and ivory doors, on patios scented with honeysuckle and light as the Alhambra.

Perhaps it wasn't the absinthe . . . he had only taken three drops. It may have been the gentian and dandelion infusion he had enjoyed on the Palace verandah. He needed to tell his dream to Indaco de Lunas, like a confession, he felt so guilty; but events conspired to prevent him from speaking to Indaco, alone, until late that night, after a memorable game of bridge between Indaco and the noble Venetian Dal Bon against Professor Carlo Lenzi Grillini of Florence and his Amazon travelling companion, Ivano.

"Two spades," Indaco bid nervously, not so much because of the points, which he had, but because he only had four spades, albeit the ace, king, queen and jack. True, that singleton with the ace of diamonds reassured him.

"Pass, bless you!" replied Lenzi Grillini, blowing smoke rings with his Macedonia Extra.

It was Dal Bon's turn, after the Tuscan irony of that blessing. With the calm he exercised to import ships full of beans or of chests of jade from China, the young Venetian aristocrat announced, "Seven spades!" He only had four points in his hand.

"But this is Pieri's coup," observed Indaco, recalling a similar declaration in Sardinia with his Sienese friend Beppe, a reckless more than great player.

Actually, Dal Bon, a new Marco Polo, was many things: shipowner, financier, merchant, traveller, always ready for humanitarian causes, and personal friend of the Saudi king and Afghani tribal chiefs. He frequented the Palace Grand Hotel to salve his beautiful young wife's depressions.

The grand slam took only a few minutes. Dal Bon had seven spades, no clubs and only the ace of hearts (his four points). Lenzi Grillini played the ace of clubs, triggering a series of trumps. It was commented that 11 Englishmen throw themselves into the Thames each year for neglecting to force the opponent to play his trumps.

Vieri wrote, "This story, too, ends in the *secche di gennaio*. Perhaps it is just as well. It began in too erotic a manner."

10

Carlo di Capri

Peasants and Owners in the Sienese Countryside

Tuscany ridges marked
by cypresses as exclamation marks.
Four limes and a cedar of Libano
for each owner-mansion.
Grandfather Edoardo is on the wrought iron armchair
in the garden. He has preferred
only violet tones in the trees
and in the bushes around the old tufa house.
With a wise slowness Olinto del Viligiardi
waters the new-born salad with a little water.
The German soldiers will never find out,
under the garden ground, the family's wealth,
bicycles and radios, silver objects and pictures
sleep quiet sleeps under salad coverings.
With the orange of the day the hoopoe came
to peck among the zigne in the bed of fossil shells.
One by one grandfather Edoardo picked them up,
oysters of Devonians, white stones
in the ochre land of Siena.
"Don't go and play with the peasants' children,"
shouts, unheard, my grandmother at the door.
From the large alley, among dusty grapes of malvasia,
Anna del Borgianni comes back home
"I will marry you, when I'm of age."
"Don't say nonsenses, you are the owners' son,
you will be never allowed to marry a peasants' daughter."

The billiard at Bar Lizza and Karl Dieter von Bötticher's scull

When Rosaluna went back in time, she often frequented friends at Bar Lizza in Siena in the sixties, but sometimes Altair was inattentive and Rosaluna had three different story beginnings. Science teaches about uncertainty, especially when dealing with time machines.

First story beginning

Bettino Memmi threw his cards up in the air and cried, "Enough!" His partner turned white and wondered what he had done wrong. Bettino was the best player at Bar Lizza. Considered legendary at poker, he turned to bridge with great ease, possibly thanks to lessons from the aspiring pharmacist, Dr. Bernardino Paradisi. Memmi's partner wracked his brain to understand how he had made a fool of himself with Memmi at the third deal, after two splendid hands . . .

"Enough!" repeated Memmi. "I have reached perfection, I'm giving up cards."

Teenagers sipping Recoaro orangeades through straws were glued to the radio broadcasting the Tour de France. They turned and stared at Memmi.

It was the early 1960s. The economic boom in Italy had not yet reached Siena. Old Nannini had made his fortune through the pastrycook's art and bottles of liqueur put down before and during the war: his pastries were the only real boom in the province. People went to Bar Nannini to ogle the barmaids: one had been voted Miss Tuscany and an elderly eminent professor of medicine had lost his head for another. However, the bar where friends met daily and spent long summer afternoons was Bar Lizza. For them, the billiard of Bar Lizza was the centre of the universe.

Second story beginning

In Bremen, Carl Dieter von Bötticher awoke in a good mood. Despite his 80 years, his memory was vivid. The previous day he had received a letter from his friend Indaco de Lunas, about to leave Siena for Sardinia, who sent regards from his friends at Bar Lizza. Indaco had sent a script for an event that took place in Russia at the end of the last war. Karl read it eagerly as it concerned him. Everything was so clear, so detailed. Karl loved the Italian language and culture. The von Bötticher family had first gone to Capri in the winter of 1930.

On 14 June 1935, Pericle Becchini, one of the sons of the pharmacist Tista, harnessed the grey pony to the gig at Laticastelli, the farm near Rapolano Terme, and set out for Siena. The wife of his dear friend from Bremen, Christian Dieter von Bötticher, a German tobacco merchant and owner of a *tabacalera* in the Reconcavo Bahiano where the smoothest cigars in the world, second only to Habanas, were made, would be arriving at the station that afternoon. Frau Marlene descended from the train in all her beauty and elegance. Accompanied by her two sons, the ailing Friedrick, the reason why the family was moving to Capri, and Karl, whose story was to cross with that of the Becchinis and Rosaluna's family.

Karl nodded and smiled at uncle Pericle and the two took an immediate liking to each other. The von Böttichers had decided to rest for a few days at the farm of their Tuscan friend, before journeying on to Capri.

On the Sunday, a lavish celebration was held in their honour. Linen table cloths were taken for a picnic on the grass near the River Ombrone. The picnic chest for important occasions, made of oak and walnut, was packed and the Indian cutlery from Jaipur, which looked like silver, was included. Vintage wines were uncorked. The Becchini girls were delighted with the handsome refined German boys who spoke reasonable Italian.

Many years later, Karl, by then Carlo di Capri, was to deceive a true Neapolitan, an Italian army official, with his perfect Capri dialect, and a forbidden dream was often to disturb the slumber of one of the girls at that picnic, by then happily married. On hot nights, Karl appeared in dreams to the young, scantily clad wife. He would take her in his arms and carry her to a hunting hut in the forest, unbutton her blouse and begin to kiss her breasts. The young woman would then always protest and Karl, with the same courteous nod he had addressed to uncle Pericle at Siena station, would say, "As you wish, the dream is yours!"

In Capri, Karl Dieter von Bötticher excelled at the gymnasium-lyceum, matriculating with an average of eight (nine in physics and philosophy), became a leader of the bands of his local peers, loved many Capri girls, corresponded courteously with the other Becchini girls (not the one of the dream) and wrote several songs in Neapolitan dialect under the name of Carlo di Capri.

The call to arms in the army of the Reich was a bolt from the blue. He managed to wangle a post as interpreter at the German command in Naples and as organiser of a brothel for officers in Trieste, but soon found himself on the Russian front with other German boys. The *Julia* Alpine troops were retreating in the bitter cold while tank battles raged. Despite shrewdness learned on Capri, typical von Bötticher determination, hiding places that nature offered him and his charm, which also worked on Russian girls, he ended up a prisoner in a Russian camp near St. Petersburg.

"Carlo!" exclaimed a voice with a Neapolitan cadence. "You're in here too!"

Third story beginning

Benito Mussolini was dressed in white the day (1937) he came to Capri and gave his nephew, Fabrizio Ciano, his first boat. It was a yacht, almost a toy, built on the shores of Lake Garda. The first time he tried to sail it, the boat overturned and Fabrizio's two German friends began to jeer in the manner of Neapolitan street urchins. Carlo and Federico were 11 and 14 at the time and were proud of the wooden scull with which they sped over the sea below their villa at Unghia Marina, the house of Julia during the reign of the emperor Tiberius. One day, while digging under some mosaics in the garden, Carlo found two Medusa heads. He mounted one on the prow of the canoe. Because of Federico's delicate health, they had been living in Capri for a number of years. They studied at the Certosa and spoke Neapolitan dialect like perfect urchins.

Carlo's full name was Karl Dieter von Bötticher, though his maternal grandfather, owner of a major tobacco company in Bremen, was called Carl, with a C. Once Bremen was the main port for the import of cigars, coffee and Bordeaux. The family company, founded in 1864, boasted 111 shops in major German cities and a tabaclera in Reconcavo Bahiano, Brazil. The von Bötticher shops sold Peterson pipes from Dublin and the best briar, porcelain and meerschaum pipes in the world.

Eleven years later, in the Russian prison camp at Pilawa on the Vistula River in Poland, Carlo remembered that canoe. He had just exchanged his elegant white tennis jersey, knitted by his mother, for a piece of bread. He was fond of the jersey, even if he had outgrown it in height, though not in width, because the first two of his five years in prison were passed in Russia, where he was reduced to skin and bone. He had not been able to remove the stains from a nose bleed that occurred when a Russian guard had struck him with a shovel. Now an elderly gentleman, Karl Dieter von Bötticher, grand master of the Freemasonry, used to tell how his guardian angel saved him on that and other occasions.

The canoe, propelled silently by two standing boys, balancing gracefully, sped to a secret place in a grotto where countess Edda Ciano Mussolini used to bathe naked, far from indiscrete eyes. For Carlo and Federico that beautiful undine was an inspiration for erotic dreams and their first sexual desires. Sometimes they would recount their first secrets of love and sex to each other in Neapolitan, so that their mother could not understand.

Once they went to Villa Ciano with Fabrizio to steal cigarettes. Edda had been invited to Villa Sole by Mafalda of Savoy and Prince Philip of Assia. The boys entered the great salon, heading for the mahogany and silver cigarette box containing Turmacs and oval Xanthias. They then smoked their first cigarettes.

The cigarette that the Neapolitan soldier offered Carlo at Pilawa was the stub of an Alfa. Carlo realised that his friend from Capri and the other three Italian prisoners had not been identified by the Russians, who thought they belonged to the German army. The Armistice had already been signed. Carlo asked to speak to the field commander and explained the situation in Russian. A few weeks later passes arrived for the Italians, who would be repatriated via Odessa. There were five passes indicating name and age but without photos. One of the five had evidently died in a previous camp. The friend from Capri and the other three, all Neapolitans, sought out Carlo. "Come with us, we have five passes and the age is the same. In Capri there are blond boys with blue eyes and you speak Neapolitan perfectly."

Freedom was at hand, but Carlo frowned. The war was nearly over and were he discovered, the life of his four friends could be endangered. He declined their kind offer. They embraced and bid each other farewell. The friends were not to see each other again for 15 years. After the end of the war, Carlo remained a further three years in the prison camp.

Anna del Borgianni

While the von Böttichers were living these events, the family of Indaco de Lunas and Veliero Edoardo Verdiano, both still close friends of Carlo Dieter von Bötticher, continued life at Pacina.

The silver Aprilia was labouring up *Strada della Massa*. Behind it, the dust of the hot summer of 1947 obscured the view. In the last few kilometres through La Pietraia and Fornaci, some peasants had run from the fields and were waving their hats from a roadside bank, or at least that's how it seemed to the only passenger in the car. In the back seat, nine-year-old Veliero Edoardo Verdiano swelled with pride. With the old chauffeur all to himself, an average mark of nine for admission to secondary school, a year ahead at school, that summer the usual trip from Siena to the farm to pass the holidays with his grandmother and uncles seemed a kind of apotheosis.

Near the Fantasticis' lemon house, in the shade of the *parata*, at the edge of the Camaianis' threshing yard and near the gate of the villa, a good number of peasants had gathered to welcome the young master back. They may have been there because the parata offered shade, out of curiosity to see the Aprilia or to enjoy the aroma of hot bread and ciaccino wafting from the oven, but the boy liked to think they were there to celebrate his return. He particularly liked to think that a certain pretty girl, few years older than he, with a wonderful complexion and almond eyes, was there for him. Admittedly, she was the daughter of a peasant family, but she had a name reminiscent of ancient aristocracy: Anna del Borgianni.

Most of the families at the farm had high sounding names: Viligiardi, Camaiani, Vanni, Maggi, Veltroni . . . and their individual names, too, were full of evocations: Clorinda, Olinto, Asmara, Virgilio, Ermellina, Argentina . . . . Several years later, red flags flying on the haystacks at harvest time would divide peasants and landowners. Share farming, which had moulded the splendid landscape of Siena, was at an end, but those struggles and political battles were always conducted fairly and with great civility: the civility of the peasants and landowners of the Tuscan countryside.

Veliero Edoardo Verdiano used to observe that just as hyphae are the intelligent network of a plant ecosystem, a forest, a meadow, passing information and defending the system from environmental extremes, so the roots of a people that has lived for generations in a place are the soul of that people and their culture, the historical memory of the land.

Mitzi

Mitzi came from Rovereto della Luna. She arrived in Siena in the spring of 1937, just before the war, and took work as a domestic help. Full-breasted and only 19, she was as strong as a stone mason.

Ten years later, a shiny, dark green international truck, made in America, was travelling the route between Osteria Bianca and Forte dei Marmi. It was a war relic bought by the family to replace the old truck "acquired" from the retreating German army. Under the tarpaulin there were four mattresses where children could be heard kicking up a rumpus. They were between five and nine years of age, two girls and three boys. Aunt Rosina could hardly control them. When Mitzi began to sing "quel mazzolin di fiori," they joined in. Then Mitzi began to whistle *Capriccio italiano*.

They were taking everything they needed for a month at the seaside. The house rented at Forte dei Marmi only had bedframes, kitchen furniture, chairs, a divan and two armchairs. The truck was loaded with pots, crockery, cutlery and sheets, besides the inevitable buckets and spades for the beach. They had been sure to include "marbles" to run on the tracks they built out of sand. These were up to ten metres long, with parabolic curves, mountains and tunnels.

Grandmother Lidia kept the children quiet with stories about trips to Forte dei Marmi when she was a girl: "I was only four. It was 1909 when father bought his first automobile. It was an event, not only for our family but for the whole village. Some had never seen a car. It was a solid white Diatto with bellows like a carriage. It seated four but could easily carry seven. The motor was started by cranking, which was hard work. There were no side doors and the probability of falling out was proportional to the speed the motor could achieve on the dreadful roads of the time. Clouds of dust rose in summer and winter as the automobile banged over the sharp stones. Animals were terrified by the noise of the motor. It was a tragedy when the car had to overtake a cart drawn by oxen. The oxen took fright and put themselves broadside in the carriageway or stampeded into the ditches. The peasants were even more afraid of the oxen, lost control and panicked. However, in a few years the animal world got used to the mechanical monsters."

"How wonderful it was to travel by car! We could go to Arcidosso to see our aunts in a few hours, whereas by train we had to change at Asciano and take a carriage from the station of Monte Amiata, which took nearly all day. We could also go to the seaside, to Forte dei Marmi! It was a long trip, but interesting. As we drove through towns, women and children would appear in the doorways and even the men would turn. The passage of a car was a relatively rare event.

After half a day's drive, we would pull into the square of Forte dei Marmi, in front of the Caffè Roma. People would crowd around. This was 1909 or 1910. There was the fort, the Pensione Idone, about 20 houses of local people and great yards full of marble slabs and blocks ready to be loaded on sailing ships that came to the rickety old wharf. There were few people like us who came to enjoy that clean sea and stretch of immaculate sand, and none of them had a car. There was a youth who had a type of motor boat and struck up a friendship with my father. They shared a passion for motors. We children liked the smell of petrol. Grandmother would make a face and mother could not stand it. It reminded us of drives through unknown countryside between long lines of trees."

Sometimes grandmother Lidia would tell of the time she saw Gabriele d'Annunzio on horseback, followed by a girl with long hair flowing in the wind. The girl was Eleonora Duse. She would tell about the players of the Italian national soccer team, Ferraris IV and Orsi, who used the same bathing establishment, and about the day at the Expo when Gabriele D'Annunzio did not come with Eleonora Duse.

"The girl in brightly coloured clothing who was with him could have been Armenian, or a gypsy or a mixture of Arab and Spanish. She had a shawl around her shoulders that was electric blue shot with pale indigo. The poet was attracted by the Art Nouveau lamps designed by Edoardo Quirino and draped the shawl over the glass. A beautiful blue light shone through. Several weeks later a parcel was delivered to the house in Rapolano. The sender was Gabriele D'Annunzio."

A few months before her second trip to Yucatan, Rosaluna was under an umbrella at Forte dei Marmi, fantasising about the stories of those times. The figure of uncle Angelo emerged clearly from the family stories. Rosaluna had only seen photos of him, a tall gentleman who came to the beach dressed meticulously in white, down his shoes. The gentleman who was approaching her now was also dressed in white and was tall and elegant.

"Carlo!" she exclaimed. Karl Dieter von Bötticher had come from Bremen to see her.

"Ciao Rosaluna," he began in perfect Italian. "I found funds for your research and your trip to Yucatan. You can prepare to depart."

11

St. Louis Blues 53100[1]

I Was Afraid

I was afraid of the sea and so didn't go
to the fiesta at Half Moon Isle.
I stayed on the wooden jetty to watch
the green boat of the Creole Garifuna.
High among the clouds flies a frigate bird,
a lonely pelican fishes between the jetties.
At five in the evening you haven't come back yet
and I'm afraid you've left me alone.
Darkness is falling down and the boat doesn't appear,
the frigate birds have gone to sleep in the bay,
now I'm alone ashore, I who feared the sea.
A small dog barking keeps me company.
The boat has come from the barrier reef,
she too has come back from the fiesta,
now I don't fear the sea any more, you are beside me,
but, another time, don't go, stand by me.

The week before her second journey to Yucatan, Rosaluna received a parcel from her grandfather's friend, Beverly, in London. It was the stormglass that Rosaluna intended to take with her. She organised her departure with Sinbad. She spoke at length with Indaco

---

[1] 53100 is the postcode of Siena.

de Lunas. She took notes. She prepared her bags: compass, camera, Winchester, medicines. She passed a few days with Altair, who would not be coming this time.

Ummi Ummi Ummi

The first thing to do was find Gamusa, the Garifuna Creole from Dangrija. Rosaluna knew that he lived at Caya Caulker and moored his boat near Martinez's bar. She had seen a photo of him from 30 years before; by now he must be over 80, but "gentlemen of fortune," their skin weathered by sun and salt, never seem to age.

The most difficult thing had been to avoid six or seven young bag snatchers in Belize City and find a boat to take them to the island before nightfall. When they reached the jetty in front of Martinez's bar, Rosaluna's attention was arrested by a group of people organising a barbecue on the beach: lobsters, enormous blackfish, a few reef fish. Several persons were up trees knocking down coconuts. The notes of a ragtime and fumes of rum wafted from the bar. Over the door hung a placard advertising "Martinez rum punch." Rosaluna recognised the dusky-skinned girl with the fluttery skirt and the décolleté that Vieri had described 30 years ago.

In the centre of the group, Gamusa had hardly changed. Only his dreadlocks were greying in places. His fine posture and complexion indicated a man of the sea. He was tossing fish entrails into the air to four frigate birds that were catching them in mid-flight. Rosaluna waxed ecstatic and nostalgic. The notes of *As time goes by* sounded in her mind with an old Sienese refrain that Vieri often used to hum:

Ummi ummi ummi
stop drinking rummy
stop drinking punch
you're such a dunce.
Drunk on Sund'y, drunk on Mond'y . . .

Either Martinez's poster was inspired by the song or the song by the ad for Martinez's rum punch. Or fundamental things are always valid, wherever one goes.

Gamusa invited Rosaluna and Sinbad to join them. He observed her with curiosity and asked if she were Italian. When she admitted she was Vieri's granddaughter, he jumped to his feet, looked her in the eyes and embraced her saying something rapidly seven or eight times in Garifuna dialect. Then he said, "Papito, my papito is no longer with us, is he not?" Papito, father, was the affectionate name Gamusa used for Vieri who was just a little older than he. He wanted news of Vieri, the family, grandfather Edoardo Quirino, how they died and about the war, Pacina, "mamita" and the girls.

"When I am in the boat at night," he added in a mixture of English and Spanish, "I am *tranquilo*, I look at Xaman-Ek and I know my papito is with me." The Caribbean commotion lasted about three seconds, like the rain.

During dinner, Gamusa explained that he was leaving for a three-day excursion in the morning, with a group of friends: some French, an Irish girl, a Swiss girl and a couple of Neapolitans. They were going to the coral archipelago of Turneffe, Half Moon Cay and the Great Blue Hole.

Before becoming art, colours and forms belong to nature. According to Lucretius, time does not exist *per se* but in the things of nature. It is time that models the *forma fluens*, the evolving fractals of shells, butterfly wings, coasts and corals. The masterpiece of time was the Blue Hole, a perfect circle, cobalt in the abyss, aquamarine towards the coral walls. There lived the great Portuguese man-of-war, the tiny trapezoid shell of van Hyning, the green, turquoise and ochre angelfish, the azure parrotfish and the midnight parrotfish.

Few days later, as they sailed south together, Rosaluna asked Gamusa whether Tina, his secret love in the days of Vieri, was Miss Fisher. Gamusa confirmed Rosaluna's suspicion, and when they sailed up the Rio Dulce from Livingstone to the Fisher's great wooden house, he preferred to spend the night on the boat.

Lady Fisher, now much older than in Rosaluna's photo, welcomed her and Sinbad in perfect style. They dined on the verandah overlooking the lake: lobsters, port and Vieri's favourite cigars. Rosaluna conversed with her late into the night and studied maps for her expeditions: one to the north on the Peten track in the direction of Tikal, one to the south to the ruins of Quirigua.

In the morning they found everything packed for them in perfect English style: two changes, drivers, provisions and plenty of fluids and ice.

During the journey, Rosaluna gradually pieced the puzzle together from Vieri's notes, what she remembered of the stories told her 30 years before and fragments of Gamusa's comments in Garifuna Creole. Errol Aurelio Guilin, Gamusa's baptismal name, was born in Dangrija in British Honduras sometime before 1920; his father was Belgian and his mamita Carib. He hardly knew his father who was over 60 at his birth; what he remembered was good and included a pair of green eyes in a dark face. Like many Europeans at the end of the 1800s, Errol Aurelio's father had a plantation in Honduras and another in Guatemala. Through business, he knew the Fishers well and the two children, Tina and Gamusa, had often played together. By the time they fell in love, they had gone in completely different directions: Gamusa had chosen the cayas of the reef, the life of a nomad fisherman at the limits of the law; Tina inherited the Fisher Company at the death of her father and was loaded with responsibilities and duties, as well as cash. Their secret, passionate, tormented love affair bloomed and wilted quickly, as things tend to do in the tropics. The memory of it remained pure and strong for both of them.

Errol Aurelio Guilin had since tried many trades and many women. He fished lobsters and later turned to the more difficult search for black coral, beyond the reef. More recently, with money made trading pearls, he bought a bigger boat for cruises and fishing, and opened a kiosk at Caulker Cay, where his 18-year-old daughter sold T-shirts that Gamusa coloured by hand when the weather was bad. The girl had a beautiful face and her father's green eyes. Her mother, Gamusa's second love, had died at 26. There were always fresh flowers near the little white cross on Delia Marine's tomb in the tiny seaside cemetery at Caulker Cay.

Gamusa produced a small pink velvet bundle from his old bag. He unfolded it and showed Rosaluna a piece of black coral, corroded by the sea. Some letters were discernible, it looked like a crossword. Many letters were illegible, but a word was clear: SATOR. To the left of the word was the symbol of the polestar, Xaman-Ek.

"Left!" thought Rosaluna, "Xaman-Ek is to the left of the magic word that indicates the site, which means there is a site east of Tikal." She brought out the maps and compass. If Vieri's sketch was correct, Palenque and Quirigua were exactly the same distance from Tikal, west and south, respectively, and the Blue Hole was the same distance east. The Blue Hole was the magic reference site, the SATOR east of Tikal. It seemed obvious to her that there was also a site the same distance north of Tikal. Sinbad pointed out that polestar means "compass rose" and that there must also be a fourth cardinal point. The fourth Maya city, thought Rosaluna, the one of the Popol Vuh, that had never been found, possibly founded by the refugees of Mayapan. However, Rosaluna knew that there was nothing in that part of Mexico, not the slightest race of a city or a ruin.

Xaman-Ek

Rosaluna opened the folder that Indaco de Lunas gave her before she left, with instructions that she was not to open until she was on her way.

This is Indaco's story.

That morning in 1518, Juan de Grijalva thought he was dreaming. Were it not for the tropical heat, the series of coral beaches, one prettier than the next, the palms, toucans and the transparent turquoise sea, he would have sworn he was in Europe. The castle overlooking the rocky coast could have been a medieval Norman or Italian fortress. He thought it advisable to give it berth rather than face the cannons that were presumably lined up behind the slits. Thus, he missed his chance to conquer the Maya city of Tulum, a year before Cortes landed in Mexico.

"That is what the historians write," added the Marquis of Grijalva, as he lounged on the patio of the Maison Marques at Valladolid, in Mexican Yucatan. "But the family documents and the notes of the brotherhood of that Sienese saint, Bernardino, tell a different story. Consider that the monks founded the convent here in Valladolid in 1552 and that my ancestor chose to retire there at an elderly age. There it was that a Sienese monk, fra' Provenzano, wrote his secret memoires. A sort of confession."

All ears were tuned: that story of sailors, monks (especially Sienese ones), family secrets and the enchanting background of Tulum had captured the attention of all present.

"That morning in 1518, Don Juan did not dare berth under the fort of Tulum, but curiosity was high. He decided to sail north along the coast and anchor in a bay about ten kilometres from Tulum. There he made a discovery that changed his life."

The Marquis of Grijalva talked late into the night. Daiquiris, piña colada, tequila and sangrita were served more than once.

Next morning, I tried to piece together the precious information from the night before and reconcile it with the complicated hypothesis sketched by Vieri.

In the document he gave Rosaluna, Indaco continued to describe his journey to Valladolid about 20 years earlier. Briefly, the manuscripts of the Marquis di Grijalva and the notes of the Sienese friar told of a sort of vortex in the bay, 20 metres from the shore. Some sailors were sucked into it while they sought respite from the heat in the water. Given up for dead, they reappeared inland a few minutes later. The vortex flowed into an underground cavern that led to the Mermaids' Cenote. In those days, the caverns had not been completely invaded by the sea, and boats from the Spanish galleon loaded with months of provisions made a long underground journey that led Don Juan de Grijalva to the discovery of the Maya underground empire, or rather that founded by the Toltec refugees from Tula led by Quetzalcoatl. The undiscovered city, the original ancient Mayapan located at 19°5 N and 91°5 W by Sienese friar and fruitlessly sought by Vieri, was underground and had been the fortune of Grijalva. The manuscripts had discontinuities in various places; either Don Juan could not or did not wish to tell all, or the monk had kept the secret of confession.

They seemed to describe underground rivers that joined the great Maya centres and opened into cenotes. Somehow the tunnels of Tikal from the Palace of Masks must have been connected with the sea. Don Juan had apparently organised an underground trade in gold and jade, unknown to the settlement of Mayapan, the city of Tulum or the King of Spain. The crew of the galleon must have been in the game and presumably swore on the Mermaids' Cenote to take the secret with them to the tomb.

By those strange coincidences that enrich life and the human history, an oath sworn by so many persons had weathered the centuries and would probably have gone into oblivion were it not for a Sienese monk, a Mexican marquis with a passion for family tradition, and a Sienese scientist, also fond of family manuscripts.

Three months after that evening in Valladolid, an expedition of archaeologists and naturalists was ready to explore the underground passages travelled by Don Juan de Grijalva. The University of Siena had contacted the German manufacturer Bruker Spectrospin, specialized in magnetic resonance instruments and also holder of a patent for small submarines for underwater expeditions. On the morning of 16 September 1980, two new bright yellow submarines descended into the depths of the great cenote of Valladolid, much to the surprise of the animals of that wondrous ecosystem, used to living in perfect solitude. Jet black fish of the limpid dark depths swam close to the porthole. One of the two submarines made a long subterranean journey.

Twenty years earlier, Jean Jacques Cousteau had taken his bathyscaphe down into the Great Blue Hole at the eastern point of the compass rose imagined 1200 years before by Maya architects. A family of manatees, the mammals of Caulker Cay so reminiscent of mermaids, approached the little yellow vessel.

The expedition found the remains of a city, but all the riches had been gone since 1518, the time of Juan de Grijalva and his Spanish crew.

Thus ended the story told by Indaco de Lunas.

The two moons of Amarilli the Second

Indaco's folder also contained a small sealed envelope on which was written "For Rosaluna – top secret." Despite the age difference between them, they had learned to confide in each other. Rosaluna read the secret diary written by Indaco with himself as protagonist.

1980: In the bay about ten kilometres south of Tulum, Indaco collected some shells on the white beach of the mermaids' cenote. They were for the malacological collection of his grandfather in the house at Pacina. Amarilli II was looking for fragments of coral along the tideline, her breasts bared to the winds of Yucatan. Indaco observed her with admiration, before reclining on the sand and closing his eyes to the sun. It was a trick that uncle Vieri had shown him, a game of colours. He remembered the words of Wolfgang von Goethe in his *Theory of Colours*: "Balancing weights and counterweights, nature oscillates this way and that, creating a here and a there, an above and a below, a before and an after, to which all manifestations in space and time are subject." He pondered on the relations in nature that bring space and time together, that give rise to form and colour, that link the observer with the object observed, that mix objective and subjective.

With his eyes closed, Indaco saw a wonderful amaranth red. When he tensed his closed lids he saw a pastel green, and when he relaxed them he saw golden orange, the colour of *Cypraea aurantium*, one of the rarest shells in the grandfather's collection.

Amarilli II called Indaco with urgency. Some manatees had appeared a few metres from the shore. The female was tummy up, suckling a pup. The young couple watched in amazement: these freshwater mammals lived in the ancient cenote and did not mind a dip in the sea. They seemed a cross between dolphins and dugongs.

The vermilion sails of Gamusa's boat appeared on the horizon. Gamusa dived in and swam towards the strange dolphins. He stroked the female and played with the pup. The animals felt his charm and followed him over and under the water, with joyful pirouettes and somersaults.

Not far from Tulum, Gamusa had found what he had long been seeking: the original notes in Maya of Alguer di Chichicastenango, the Indian monk, follower of Bartolomeo de Las Casas, whose writings had instructed and enchanted Emiliano Zapata, before the Mexican revolution.

The water in the cenote was cool and green: a circular lake about 15 metres deep. They dived in. There were caves towards the bottom and light from the other side was shining through some of them. One cavern broadened after about a hundred metres where it connected to the sea, off a white beach covered in tiny yellow and green shells. The fresh water mixed with the salt, and schools of colourful reef fish hovered and darted in the opening. A marine cenote had formed in the sea and Indaco finally realised that the cenotes were perfectly round because they were former coral atolls.

What amazed them most of all was to meet a pair of manatees. The large sirenids approached with great familiarity. They lived there, between sea and cenote, and had left their pups somewhere safe in the meanders of the caves. The legend of the mermaids was at hand. The female manatee has breasts and a mermaid's tail; she is more than two metres long and swims with sinuous grace. It was not clear where they found other members of their species to mate with. Perhaps the caves led to other cenotes which are common in Yucatan, or it may have been a small family tribe. They bid farewell to the two manatees, who accompanied them back to the cenote, and left Tulum loaded with wonder and colours.

Conversation of the two moons between Amarilli II and Indaco de Lunas or between the manatee and Indaco, or between Indaco and Egle.

Indaco: What a magic night! The mist has dissolved leaving the starry sky with two yellow slits that are watching us.

Egle: I only see one.

Indaco: No, don't you see? There are two moons!

Egle: I know, Indaco. I can only see one, but I'm sure that if I close my eyes I can see two like you.

(Indaco de Lunas turns slowly towards the girl seated behind him in the small boat. He looks at her intensely. The night is bright and the rays of the two moons light her face and her closed eyes. They seem to be there just for this.)

Indaco: It's incredible. It's the first time I've ever felt so at ease with a woman.

Egle (*smiling, still with her eyes shut*): I see you Indaco, I can see you anyway. There are still two moons, aren't there? They are there in front of us. They light our faces. We are protagonists of this magic night, aren't we, dear sailor?

Indaco: Don't they seem to be doors in the sky? One for me and one for you. Why don't we try to open them together?

(Egle, eyes still shut, squeezes Indaco's hand and smiles. For him it is a moment of great intensity.)

Indaco: Do you hear this music too? It's Mozart, isn't it?

Egle: Yes, I know it. Beautiful! It is taking us towards the two moons. My father used to say that when Venetians are tired of the authorities, they go to magic places. They open the doors at the bottom of the courtyards and depart forever to beautiful places and other stories.

Indaco: We are flying. I feel so light.

Egle: The moons are closer and closer!

Indaco: Come! Let's go in together!

Egle: I cannot, Indaco. We belong to different worlds. My moon is different from your two moons.

Indaco: I won't let go of your hand.

Egle: You'll lose me anyway. But one day, as in the past, we will belong to the same world and then we'll open those doors together.

(Egle opens her eyes and the world comes back to normal. There is only one moon in the sky, and Indaco is still there holding her hand.)

Egle: See, Indaco, for me there is only one moon, whereas for you there are still two, aren't there?

Indaco: How can it be? If we belong to different worlds, why am I so certain that I've always known you?

Egle: Our paths crossed a long time ago.

(Indaco looks sad. He takes up the oars and rows in the direction of the small jetty. Egle buttons up her jacket, leans her head gently on his back and starts to dream.Again she sees two moons, and she smiles as she dreams.)

Thus ended the secret diary of Indaco de Lunas.

Calculations of a mathematician born in St. Louis

Rosaluna remembered the words of the Indian woman in the church at Chichicastenango. She saw a large flock of birds that suddenly divided in two. One flock flew off in the direction of the jungle of Peten, the other towards the sea.

She immediately changed her plans. The expedition would divide in two. The first group, led by Gamusa, would obtain objects and information on the Popol Vuh in a long journey overland to Quirigua, Tikal and Palenque.

"We know enough about the Blue Hole from Indaco's descriptions and the work of the French scientist." Gamusa agreed, despite himself: he was always ready for a trip to the Blue Hole.

"The other group in the boat with me and Sinbad will collect Indaco's second yellow submarine near Tulum. We will sail north."

"Suerte!" said Gamusa.

"See you at Martinez's next week," said Rosaluna.

On board, Rosaluna and Sinbad had time to talk and think. Why had the Popol Vuh not been found in the underground city? Why didn't the monk mention this sacred book? Why had Indaco denied any sacred importance of the city? There were no idols, votive objects, scriptures or even astronomical calculations, at which the Maya were masters. Masters, calculations . . . . Rosaluna remembered the last words of the Maya woman and dialled Italy: 335-6592114.

"Elisa? It's Rosaluna, I'm at sea off Tulum. I have here the longitude and latitude that you know. Could there be errors?

"Funny you should ask. I was thinking the same thing these days as I exercised Altair," replied the Italian mathematician born in St. Louis.

"Actually, the area has highly magnetic rocks, which you should be able to detect with your stormglass. A strong magnetic component could lead to an error in longitude. I tried to calculate it again on the basis of what I know and I got 89°3 instead of 91°5.

Thanks Elisa. A hug also from Sinbad!"

Rosaluna was excited and she magnetised the crew. Twenty-four hours later the submarine was already at the Mermaids' Cenote of Tulum. It had been kept in excellent state by a guardian in uniform who had received a wage to lubricate and keep the submarine in operation. From the cenote, they travelled south to Bacalar lagoon by underground rivers and then north into Mexico as far as 91°5. Here the stormglass indicated a magnetic storm right where the underground river divided, one branch in an easterly direction. Many animals, too, migrate guided by tiny magnets in their body and somehow related to the supramolecular structure of water.

"Right to 89°3!" ordered Rosaluna.

The volumes of the Popol Vuh and four wonderful jade idols in colours ranging from dark green to black, pale green and pale blue tinged with pink appeared to Sinbad and Rosaluna in a great underground hall. They were in a specially carved niche, in a cabinet made of jade leaf. "The jade guardians (Lari)!" thought Rosaluna. The sacred book contained many passages that taught how to live in harmony with nature instead of manipulating or dominating her. Two phrases were almost identical to those written by the Indian poet Rabindranath Tagore: "Do we not know that the appearance of the seed belies its true nature? If we do chemical analysis we find carbon, proteins and many other substances, but not the idea of a tree and its leaves" and "Man has lost his interior perspective, he measures his stature against matter and not in relation to his vital link with the infinite; he measures his activity on the basis of his hurried life not on the basis of the serenity of perfection or the calm of the starry sky, the unceasing rhythm of creation." According to the book, the seed already encloses the wonder of life, the potential to evolve, and beauty.

At the end of the week, Rosaluna and Sinbad celebrated with Belisario Martinez's rum punch. Gamusa arrived later. He had not limited himself to collecting material at Tikal and Palenque useful for deciphering the Popol Vuh, but he had also deciphered the figures on the stele of Quirigua and had followed another path. From Tikal he had sailed up Rio de la Pasion via Uxumacinta to Bonampak and thence to Palenque. He saw the fresco that indicates that the roads are underground because the man is under the quetzal and astride a turtle. Strangely, the turtle is yellow. He saw a sacred place in the forest of Lacandona where the Zapatist community lives. He saw the tombs of Yaxitlan and the drawings on the bastions of Orion. He encountered howling monkeys and jaguars.

Now he was tired and was going off to dream. Rosaluna embraced him tenderly. She took Sinbad by the hand and they too went off to dream.

The elmwood table

The great table of elmwood at Pacina, Castelnuovo Berardenga, 53100 Siena, was designed and made by Edoardo Quirino, grandfather of Veliero Edoardo Verdiano.

The special day began with a soak in the thermal waters at Rapolano. The local council had named a street after the inventor: Via Edoardo Quirino. Among the reasons were his inventions in the field of acetylene lighting used in the streets of many Italian cities and his contribution to the construction of the first radio by Guglielmo Marconi: the patent was for a new type of condensor. In the old house at the corner of the Piazzone and the main road to Rapolano there are still some prototypes of radios built by Edoardo Quirino. In those days, friends would visit in summer to listen to Radio Toulouse and talk about science. There was an art nouveau lamp draped with a splendid azure-indigo silk shawl.

The table at Pacina had seen much history. In the seventies, the first Italian environmental movements were founded around it; at other times it hosted meetings of scientist friends: Nobel laureates Wilkinson, Bloch, Prigogine and Wütrich had brainstormed on innovative topics of physical chemistry. Odum, Daly and Georgescu-Roegen had also taken part in discussions around the great table.

In the afternoon the *Carro di Tespi* arrived from Sant'Angelo d'Ischia: Giorgio, Iorio's family and the legendary Crescenzo put on a comedy by De Filippo. The prize-winning fireworks company of Sant'Angelo came with them in the person of Mario, who was responsible for special effects that night. There was also maestro Mimmo Abriola with his musical instruments. On their arrival, there was a moment of emotion as they embraced Carl Dieter von Bötticher from Bremen and/or Capri, who had reached Pacina with his grandson Giovanni. Mimmo played *Era di Maggio* for the occasion.

At dusk they visited the parish church with its circular belfry, built before the year 1000, taking cherry blossoms to the family chapel for the bella Elena, Edoardo Quirino's grandfather and uncle Vieri. Unusually for him, Indaco de Lunas was visibly moved, in spite of his English aplomb. Music: *New Orleans Function*.

Before dinner, among applause from the guests, a great colourful balloon ascended in flight, taking Elisa, Giovanna and Rosaluna, the organisers, for an excursion over the Chianti. Music: *Mahogany Hall Stomp*.

The party became lively with the arrival of Gianna Nannini, who sat down at the piano to play and sing with Mimmo Abriola.

The sculptor Pietro Cascella, a dear friend of the grandfather, whose birthday he shared, arrived with Cordelia von den Steinen, bringing two gifts from Castle of Verrucola in Lunigiana, where they lived. One was a Luni woman in travertine, "one of the group" by Pietro; the other a stone bowl from Nemrut Dagi.

Under the cowl of the great fireplace, where she used to hang a rose-coloured stocking of bella Elena at Epiphany, Rosaluna found a gift from the bird nation: a beautiful pearl necklace. Music: *Mood Indigo*.

The chests of Gaspar, Melchior and Baldassar were open near the entrance. The first contained an illuminated original of the Koran from the desert libraries, a gift from Rosario Simone. The second displayed the book of the Popol Vuh that Rosaluna had brought back with her from Yucatan. The third was a recent gift, the *Canticle of All Creatures*, posted by Francesco from the route to the archipelago of 11 islands. The folder also contained the calculation of the latitude and longitude used to discover the site of the subterranean Maya city and the books of the Popol Vuh.

A letter with a brightly coloured Guatamalan postage stamp arrived from Tina Fisher. Her aches and pains had prevented her from undertaking the voyage. She wrote that Gamusa had been away for many months. He was often at sea on the *Frigate*, lending his pirate services to Greenpeace against the dastardly oil companies. Real piracy, she commented, was carried out within the paradigm of democratic rules, behind the scenes. Tina also had news of father Norberg of St. Louis, who in distant 1967 had disappeared to join Che Guevara. This was confirmed by Ernesto Cardenal, the Jesuit poet and revolutionary minister of Nicaragua, a friend of Indaco's, who had come for the occasion. Music: *Caminito* by Carlos Gardel.

An old gramophone alternated records of Bix Beiderbecke and Louis Armstrong (La vie en rose). At home that evening *Saint Louis Blues* was played for Elisa, who was born in that city.

Altair had the place of honour and a large silver platter of oats. Proud head of the table, he munched the grains with a grinding sound.

Veliero Edoardo Verdiano had taken his friends from Bar Lizza aside: a certain Paolo from Paris, a Mario from London, and Indaco who was in a particularly gay and open mood. Mario del Viligiardi had borrowed the costume of the "quattro di Balia" that he wore on the chariot during the Palio. Seated at the great elm table, near his friend Geppetto, a fine cabinet maker in Via delle Sperandie, he seemed a real medieval gentleman, enhanced by his myopic gaze, lost in the past.

Rosaluna, on Sinbad's arm, was dressed so simply and elegantly that Toni Thorimbert could not resist a few photos.

Invitations had been sent to all corners of the earth: for the day when the ancient cherry blossomed, their presence was desired at the great elmwood table in Pacina. Toasts with syrah were assured. First of April 2007, Pacina, 53100 Siena, Tuscany. Music: *Cielito Lindo.*

The guests ate, drank and conversed happily. All enjoyed themselves. There was only a pretty girl dressed in turquoise percale who seemed slightly ill at ease: her dress was too tight around the thighs. A slight tearing sound was heard as the percale gave way. One of the guests, surely under the influence of alcohol, swears to have seen the twin tails of a mermaid under the table.